Nanotechnology Law

Nanotechnology Law

Best Practices

Lorna Brazell

Wolters Kluwer
Law & Business

Published by:
Kluwer Law International
PO Box 316
2400 AH Alphen aan den Rijn
The Netherlands
Website: www.kluwerlaw.com

Sold and distributed in North, Central and South America by:
Aspen Publishers, Inc.
7201 McKinney Circle
Frederick, MD 21704
United States of America
Email: customer.service@aspenpublishers.com

Sold and distributed in all other countries by:
Turpin Distribution Services Ltd.
Stratton Business Park
Pegasus Drive, Biggleswade
Bedfordshire SG18 8TQ
United Kingdom
Email: kluwerlaw@turpin-distribution.com

Printed on acid-free paper.

ISBN 978-90-411-3826-2

Printed and Bound by CPI Group (UK) Ltd, Croydon, CR0 4YY.

Table of Contents

List of Abbreviations

ADME	Absorption, Distribution, Metabolism and Excretion
ANZF Code	Australia New Zealand Food Standards Code
ASTM	ASTM International
BSE	Bovine Spongiform Encephalytis (Mad Cow Disease)
C&L Inventory	Classification and Labelling Inventory (of the European Union)
CASG Nano	REACH Competent Authorities Sub-group on Nanomaterials
CEN	European Standards Centre
CEPA	Canadian Environmental Protection Act
CHMP	European Medicines Agency's Committee on Medicinal Products for Human Use
CLP	Classification, Labelling and Packaging (of Chemicals)
CMR	Carcinogenic, Mutagenic or Toxic for Reproduction
CNT	Carbon Nanotube
CSCL	Japanese Chemical Substances Control Law
DAR	Draft Assessment Report
DDP	Dossier Development Plan
DG	Directorate General (of the European Commission)
DNA	Dioxyribonucleic Acid
DSL	Canadian Domestic Substances List
ECHA	European Chemicals Agency
EEA	European Economic Area
EFSA	European Food Safety Agency
EINECS	European Inventory of Existing Commercial Chemical Substances
EMA	European Medicines Agency

EPA	US Environment Protection Agency
EPC	European Patent Convention
EPO	European Patent Office
ETSI	European Telecommunications Standards Institution
EU	European Union
FAO	United Nations Food and Agriculture Organization
FDA	US Food and Drug Administration
FIFRA	US Federal Insecticide, Fungicide and Rodenticide Act
FRAND	Fair, Reasonable and Non-discriminatory
FSANZ	Food Standards Australia New Zealand
FTC	US Federal Trade Commission
GHS	Globally Harmonised System of Classification and Labelling of Chemicals
GLP	Good Laboratory Practice
GMO	Genetically Modified Organism
GRAS	Generally Recognized as Safe
ICCM	International Conference on Chemicals Management
IFCS	Intergovernmental Forum on Chemical Safety
IOMC	Inter-Organization Programme for the Sound Management of Chemicals
IP	Intellectual Property
IPPC	Integrated Pollution Prevention and Control
IPR	Intellectual Property Right
IRGC	International Risk Governance Council
ISO	International Standards Organization
IUPAC	International Union of Pure and Applied Chemistry
IUR	Inventory Update Reporting
JEDEC	Joint Electron Device Engineering Council (now known as the JEDEC Solid State Technology Association)
JRC	Joint Research Centre (of the European Commission)
LED	Light Emitting Diode
MA	Marketing Authorization
METI	Japanese Ministry of Economy, Trade and Industry
MHLW	Japanese Ministry of Health, Labour and Welfare
MHRA	UK Medicines and Healthcare Products Regulatory Agency
MPEG	Moving Picture Experts Group
MSDS	Material Safety Data Sheet
MWCNT	Multi-walled Carbon Nanotube
NICNAS	Australian National Industrial Chemicals Notification and Assessment Scheme
NIOSH	US National Institute for Occupational Safety and Health
NMSP	US Nanoscale Materials Stewardship Program
OECD	Organisation for Economic Co-operation and Development
OIE	World Organisation for Animal Health
PCB	Polychlorinated Biphenyl

PCT	Patent Cooperation Treaty
PM	Particulate Matter
PMN	US Premanufacture Notice
RAE	UK Royal Academy of Engineering
RAPEX	Rapid [Product Safety] Information System
REACH	European Regulation on the Registration, Evaluation, Authorization and Restriction of Chemicals
ROHS	European Regulation on the Restriction on Use of Hazardous Substances
SAICM	Strategic Approach to International Chemicals Management
SARS	Severe Acute Respiratory Syndrome
SCCP	European Scientific Committee for Consumer Products
SCCS	European Scientific Committee on Consumer Safety
SCENIHR	European Scientific Committee on Emerging and Newly Identified Health Risks
SDRAM	Synchronous Dynamic Random Access Memory
SDS	Safety Data Sheets
SIEF	Substance Information Exchange Forum
SNUR	Significant New Use Rule
SPS Agreement	Agreement on Sanitary and Phytosanitary Measures
SVHC	Substance of Very High Concern
SWA	Safe Work Australia
TRIPs Agreement	Agreement on Trade-Related Aspects of Intellectual Property
TSCA	US Toxic Substances Control Act
TTBER	European Technology Transfer Block Exemption Regulation
UN	United Nations
UNEP	United Nations Environment Programme
UNESCO	United Nations Education, Science and Culture Organization
UNIDO	United Nations Industrial Development Organisation
UNITAR	United Nations Institute for Training and Research
USPTO	United States Patent and Trademark Office
UV	Ultraviolet
VCR	Video cassette recorder
WHO	World Health Organization
WIPO	World Intellectual Property Organization
WPMN	Working Party on Manufactured Nanomaterials
WPN	Working Party on Nanotechnology
WTO	World Trade Organization

Preface

It has become increasingly difficult over the last few years for anyone interested in new nanomaterials and products – and how is it possible not to be interested in such intrinsically fascinating technologies? – to ignore the cacophony of debate at national, European and international level over the question of whether, and to what extent, existing laws and regulations apply to nanotechnology. Conferences, working groups, technical committees and learned institutions of all kinds have considered the issues and opined, but the state of scientific knowledge has generally been considered too immature for any real consensus to emerge. Further, although the number of consumer products reaching the market which incorporated any form of nanomaterial was steadily growing, the number any consumer might realistically expect to encounter was still relatively low. Accordingly, research and development have ploughed ahead but the legal and regulatory context for the fruits of that work has been left almost untouched.

In the last two years, the question of what form any such laws and regulations should take has begun to come to the fore with the introduction in Europe and Australia of nano-specific regulation and the USA looking likely to follow suit in the near future. The OECD's Sponsorship Programme, a broad programme of research into the properties of particular nanomaterials being carried out collaboratively between a number of OECD Member States and observers, which is expected to generate some of the fundamental knowledge necessary for rational regulation, is due to report in 2011. It is therefore timely to review the various sectors in which nano-enabled products are expected to be used, the regulation which has applied in the past and the changes which are being proposed to address the potential risks to which nanomaterials are feared to give rise.

Although the book is intended for a non-technical audience, it is impossible to cover a topic in which the technical issues are the crux of the debate without some technical material. I have attempted to explain as clearly as possible the critical issues around nanotechnology – including the basic parameters as to what is and is not properly so called. But where health and environmental impacts are concerned, I have not given any detailed explanation, leaving it to the interested reader to follow up the science through the cited publications. In any case, this particular wheel does not need re-inventing: a number of excellent reports including that of the House of Lords Select Committee on Nanotechnology in Food, the Opinions of the European Commission's Scientific Committee on Emerging and Newly Identified Health Risks and that of the Royal Society/Royal Academy of Engineering, do the topic justice.

In writing, the main problem was not lack of material but how to synthesize an excess of source material including guidance, notices, commentary and draft regulation from all corners of the globe. It is almost inevitable that something will have come out between writing and publication which would have been well worthy of comment; but this would have been true whenever the book was written, unless the topic were left for another decade or more until all of the questions have been answered. This text covers the law, including draft new European regulations on Biocides and Novel Foods, as at 18 October 2011, although the treatment of the European Commission's Recommendation on a definition of the term 'nanomaterial' which came out on that date is necessarily rather brief.

Chapter 1

Nanotechnology in the Twenty-First Century

1.1 INTRODUCTION

In the last few years, the term 'nanotechnology' has become fashionable and the subject of passionate debate as to how the subject should be handled through law and regulation. Nanotechnology is discussed as an entirely new development with the potential equally to bring valuable new benefits and also to engender unknown and potentially very serious new hazards, both for human health and for the environment. But in the same period, the average consumer's chances of actually encountering any product using nanotechnology were extremely low. Until very recent years, nanotechnology has been largely confined to the laboratory with both its potential benefits and its potential risks still substantially speculative rather than actual.

This is about to change. The results of the last decade or so of research are now reaching commercial and industrial reality, and accordingly the need to address any legal and regulatory issues raised by these new products is becoming urgent. A political imperative to be seen to be managing any new risks arising has developed considerable momentum. Yet there is very little agreement as to what, precisely, needs to be done. The state of scientific knowledge is still far from complete, and the evidence for both benefits and risks will only emerge over the next decade or more. Regulators in all of the industrialized countries are grappling with the question of how best to minimize any exposure to a barely identified risk, and at the same time minimize any unnecessary burden on researchers and industry from divergent regulatory approaches around the globe.

The result is a complex and rapidly changing legal landscape. Regulatory instruments addressing nanotechnology in industrial sectors from chemicals to food and waste are now being adopted at an increasingly rapid rate, and in many cases expressly acknowledge that as the knowledge base grows the regulations may need to be adapted accordingly. For several years hence, attempting to grasp the legal and regulatory situation may only become more difficult. Accordingly, this book attempts to set out how the current debate has arisen, how the existing law deals with the issues arising around nanotechnology in areas such as patent rights and manufacturers' liability, the nature of the discussions taking place in various international organizations in order to achieve some global consensus at least at the level of principle, and the present regulatory position in Europe and selected industrialized countries.

1.2 DEFINING NANOTECHNOLOGY

1.2.1 Basic Concepts

The value of the debate over nanotechnology is undermined by a very real lack of consensus as to exactly what is being talked about when the term is used. As will be explained below, the term itself is ill-suited to the subject matter, while major uncertainties remain even as to the scientific understanding of the subject. A large proportion of the submissions to the 2009 scientific hearing held by the European Commission on Risk Assessment of Nanotechnology focused on the lack of data: as to properties of nanomaterials, methods of accurately detecting nanomaterials, means for distinguishing naturally occurring nanomaterials from engineered ones and so on.[1] A further Safety for Success meeting convened by the Commission in April 2011 reiterated many of the same points.

This book cannot directly address those uncertainties, but this section aims to ensure that there should be no uncertainty what the technologies are that it does address.

The term *nano* is a measurement of size. A nanometre (nm) is a millionth of a millimetre – far too small to be perceived by the eye, and little larger than the individual atoms and molecules which make up all matter. A nanometre is normally illustrated as being about 1/50,000th the width of a human hair, or 1/100,000th the thickness of a sheet of office paper.

The scientific system of nomenclature for classifying sizes generally operates by divisions of 100: thus, a centimetre is 1/100th of a metre; a micron or micrometre (μm) is 1/100th of a centimetre; and a nanometre is 1/100th of a micron. Accordingly, the nanoscale is commonly taken to refer to matter at dimensions between 0.1 and 100 nm since the upper limit of this range – 100 nm – in fact equates to 0.1 μm. Structures 500 nm across or larger are equally or more conveniently described as being on the microscale. There is ample scope for argument as to whether structures 100–500 nm in scale fall within the meaning of nanoscale or not.

1. *Submissions scientific hearing on nanotechnology* available at <http://ec.europa.eu/health/ph_risk/committees/04_scenihr/docs/nano_submissions.pdf>.

So much for 'nano'. The word 'technology' means the applied arts: not pure scientific theories, but the practical applications to which they can be put. Technology in the human-scale world means all of those techniques and devices which have been developed over the centuries in order to manipulate the world. This meaning is equally applicable when the prefix 'nano' is applied: semantically, nanotechnology is the collection of new techniques and devices which are being developed to manipulate matter on the nanoscale. In this sense, nanotechnology is a toolkit which can be applied to every field of science and technology: not any particular product or material, or even a separate area of study in its own right, but the collection of background knowledge which enables nanoscale products and materials to be created.

On one view, therefore, nanotechnology is a body of knowledge which in itself requires no regulation at all.

1.2.2 Common Parlance

In the majority of cases, however, when the term nanotechnology is used it is not being used to refer to this body of knowledge but to the products – nanoscale materials and devices – which it has enabled or may enable to be made.

1.2.3 Definitional Issues

Although the nanoscale is a precisely defined realm as a matter purely of mathematics, scale alone is less helpful when it comes to considering the properties of materials. At sizes of 0.1–100 nm the physical, chemical and biological properties of materials can differ in fundamental and often valuable ways from the properties of individual atoms and molecules or bulk matter. For instance, effects predicted by quantum mechanics are observed. But the upper limit of 100 nm is effectively arbitrary from this perspective: the European Union's Scientific Committee on Emerging and Newly Identified Health Risks (SCENIHR) concluded in its 2010 Opinion that there is no scientific evidence supporting it or any other single upper limit.[2] Although materials at scale of a micron or above are highly unlikely to evince the new properties shown at the nanoscale, some materials and structures of intermediate size – 2–300 nm for instance – do also show such properties. There is therefore an obvious argument for any definition of the term 'nanotechnology' to include some materials and structures up to this size.

It must be borne in mind, however, that not all nanoscale materials possess properties which are any different from the properties which the same chemical substance displays in its bulk form. For instance, properties such as chemical reactivity tend to relate to surface area and accordingly increase at nanoscale

2. SCENIHR Opinion on a Scientific basis for the definition of the term 'nanomaterial', December 2010, at <http://ec.europa.eu/health/scientific_committees/emerging/docs/scenihr_o_032.pdf>.

since the surface area presented by a given mass of substance in the form of nanoparticles is very much larger than that presented by the same mass in the form of particles of diameter 1 micron or larger. But this increased reactivity is only a linear extrapolation of the bulk scale reactivity; it is predictable without invoking any non-classical science.

Further, nanotechnology in the sense of technology being developed at the nanoscale is not new, as such. Electronics companies have been working with molecules and structures on the nanoscale for many years without that work being classified as a new 'nanotechnology'.[3] The structures which make up a silicon chip, for use in a computer or as part of a light emitting diode (LED) for lighting applications, are in fact on the nanoscale, but being comprised within a solid structure (and incapable of existing separately from that structure) are not considered to be included when nanotechnology in the new sense is discussed. There have never been calls for novel regulation of the electronics industry merely because of the ever-shrinking size of chip features. Similarly biotechnology research has been focused on manipulating matter – gene sequences, proteins and cells – on the nanoscale since the 1970s, but because the resulting products owe their properties to their complex chemical make-up and conformation rather than their size, they are not generally considered to fall under the heading of nanotechnology as such. The regulatory regime for medicines has been amended to include biological molecules in addition to small molecules (whose physiological effects are due solely to their chemical composition), but the size of the molecules concerned was not a relevant issue in those revisions.

In addition, nanoparticles produced through human activity have also been around for many years.[4] Many traditional industrial processes such as combustion, welding, explosions and so on are likely to produce at least some nanoparticles as unintended by-products. Data about the amount of nanomaterial from such processes has not been collected, but it is likely that particles which fall in the nanoscale are only a small proportion of the total dust produced. In the absence of any specific evidence of harm resulting from that component which is nanoscale, and given that such processes are already conducted under controlled conditions, there is no pressing reason to add further regulation.

1.2.4 What Is To Be Regulated

Thus, established nanoscale technologies are not the focus of the current debate and any regulatory definition of nanotechnology need not include them. Nor are they the subject matter of this book. Instead, the category to be regulated when

---

3. Although the term nanotechnology was in fact coined in the context of semiconductor fabrication techniques by Norio Taniguchi of the Tokyo University of Science in 1974, it was not adopted by the industry.
4. Those produced by natural activities such as erosion or volcanic eruption will have been around for as long as the planet.

regulation of nanotechnology is proposed is generally limited to purposely made – sometimes described as 'engineered' or 'manufactured' – nanomaterials, rather than those arising as by-products. (The term engineered seems the better option here, since literally speaking all non-natural nanoparticles must have been made by some human process.)

These new materials exploit the novel properties which otherwise familiar chemical substances exhibit when those substances are structured at the nanoscale. The most familiar example is probably that of carbon nanotubes. Carbon is one of the most commonly occurring elements and familiar in many forms: carbon dioxide in the air; charcoal as fuel (and soot as the by-product of burning coal or charcoal); and even in its mineral forms of graphite (used as the lead in pencils) and diamond. All of these 'bulk' forms of carbon are well understood and (subject to the debate on global warming) appropriately regulated by existing legislation. Carbon nanotubes, however, and the even more exotic carbon 'buckyballs',[5] have entirely different properties as to their strength, flexibility, toxicity and so on. They therefore offer new possibilities for manufacturing – lightweight but very strong and elastic structures – but also raise new concerns as to their potential impact on health and the environment.

It is this category of new forms of existing substances, designed to exploit the special properties of matter on the nanoscale, which are referred to as nanomaterials. Nanomaterials are already a commercial reality; the principal component of the regulatory debate is the question of how to maximize the benefits they may offer while also limiting the risk from unforeseen hazards.

Some existing products may, nevertheless, be affected. Carbon black is a nanoparticulate form of carbon which has been used in industrial production for over a century. In 1915 it was introduced as a reinforcing agent for the production of car tyres and today the annual production exceeds 10 million tonnes, still mostly for car tyres. Other early nanomaterials were fumed silica, a form of silicon dioxide (SiO_2), and later on also titanium dioxide (TiO_2) and zinc oxide (ZnO). These four nanomaterials, together with the more recently introduced silver nanoparticles, are produced in the largest quantities and represent the main volume of what is introduced into current consumer products.

The other component referred to by the term nanotechnology is the category of devices and structures which may be created on the nanoscale. Once again there is potential for lack of clarity here: developments such as the first synthetic self-replicating bacterium[6] are not the issue, although they may be as small as 300 nm and obviously do raise ethical and regulatory issues. That area of research is referred to as synthetic biology. Rather, when talking of nanotechnologies the devices usually in contemplation are nanoscale engines, tools and robots – mechanical,

5. Hollow spheres constructed solely of carbon atoms. The name 'buckyball' is an abbreviation of the full name, buckminsterfullerene, which was given to these structures in homage to Richard Buckminster Fuller, whose geodesic domes they were thought to resemble.
6. *First Self-Replicating Synthetic Bacterial Cell* Press release, J Craig Venter Institute, 20 May 2010.

chemical and electrical engineering on the very small scale – which today remain for the most part research projects or science fiction, but which may reach the public domain in the next 50 years or so.

It should be obvious at this stage that the task of 'regulating nanotechnology' in the broad terms in which it is often discussed, is hopelessly ill-defined. Should regulators be considering materials or machines or even techniques? Is one set of regulation suitable or will multiple different approaches be needed? If nanomaterials are to be regulated separately from bulk materials, which materials are to be included and how will they be defined for legal clarity and certainty? And further, given that the study of matter on the nanoscale is still immature relative to many other areas of science, what testing and assessment methods should be used since those methods used for traditional chemicals and materials are not necessarily effective when applied to nanomaterials? Clearly, no regulation can be effective unless it is possible to assess compliance routinely and accurately. But there is a very real possibility that when working with materials on such incredibly small scale the testing processes themselves may produce artefacts or may affect the material being tested, so as to render the results unreliable.

1.2.5 Potential Parameters

The preliminary question of defining nanotechnology or nanomaterials for regulatory purposes has therefore become intensely problematic. It would be appealingly simple to take scale as the determining characteristic: anything on or around 0.1–100 nm either as to its external dimension or internal structure (to capture aggregates and agglomerates of smaller particles, which may have the potential to revert to primary particles in biological systems or nanoparticles 'swept up' by larger particles in the air or in liquids but which may later be released) ought, on its face, to qualify as a nanomaterial. It is true that particles on this scale will be distributed within the body or in ecosystems, differently from the same material in the form of larger particles, just because of their size. But in practice such an approach would capture a vast number of materials which display no properties on this scale which they do not also display when used comprising particles on a larger 'bulk' scale. These scale-invariant materials raise no additional issues for regulation, and to include them in a one-size-fits-all definition would increase the regulatory burden on both regulators and industry, and potentially restrict the range of products brought to market, when there is no valid scientific reason to do so. However, not all materials which show altered properties or behaviour at small scale do so at any given threshold of size: the selection of (say) 100 nm as the upper limit risks excluding materials which either as particles or structures 200, 300 or 400 nm in size do show the changed characteristics which are broadly associated with the nanoscale.

The first solution which one might instinctively look for is an alternative one-size-fits-all parameter which would capture those nanoscale materials and structures which exhibit different properties from their bulk equivalents. The parameter of volume specific surface area has been considered, but proved inadequate to capture all known materials. For example, it can only be measured in respect of dry powders or solids, not materials in suspension in liquids. Nevertheless, SCENIHR's 2010 Opinion concluded that for materials in dry solid and powder forms a volume specific surface area above 60 m^2/cm^2 should be indicative of a nanomaterial.[7]

The surface properties of the particles themselves cannot be the sole factor, however, since many nanoparticles will be coated or otherwise have surface modifications which will affect key aspects of their behaviour, such as solubility in water or oil,[8] in a given context. Solubility is an important property so far as assessment of toxicity in living systems is concerned, since once a particle dissolves it is no longer important what size it was in solid form. The particle size for soluble substances is unlikely to affect toxicity as such, so knowledge about the bulk material can be applied. The 'raw' properties may therefore not be a meaningful guide to the potential toxicity or otherwise of the particles as they would actually be used.

Particle size distribution is a further complicating factor: many fine powders or dusts will contain a small proportion of nanoscale particles, but this cannot be treated as a basis for classifying all powders and dusts as nanomaterials. Similarly, most materials will have some surface structure that is nanoscale, but this will only rarely affect the behaviour of the material.

SCENIHR concluded that there was no single property which could be universally applicable as a scientific criterion for defining all nanomaterials, but recognized that for discussion purposes size with an upper limit of 100 nm is most commonly used. The Committee commented that a differentiated approach based on a relatively high upper threshold with the lower threshold treated as the more critical may be needed, with modifications of the definition for particular circumstances.

What, then, are regulators to do? At one extreme, regulators could require each and every material which is proposed to be used at a particle size below (say) 500 nm to be thoroughly tested for its potential short- and long-term effects both on human health and the environment. Such an approach would ensure that when finally approved for use the risk of any adverse consequences would be exceptionally low. It would resemble an expanded version of the pre-market testing which currently is applied only to pharmaceuticals and pesticides: expanded in that there is no requirement at present for pharmaceutical manufacturers to study the effect of their products on the environment (save in that by-products from manufacture fall within the general law on waste). But it is notorious that the testing regime applied to pharmaceuticals holds up commercialization of these materials for years if not decades, and estimates of cost range between EUR 500 and 800 m per product. These costs can be borne because of the very high profit margins which new drugs can command. Given that most new nanomaterials

---

7. See *supra* n. 2.
8. Although particle size may still be relevant for dissolution rate and hence rate of absorption.

would be destined for goods far less profitable than pharmaceuticals, imposing such stringent testing might make many candidates simply too expensive to develop for many applications.

At the other extreme, regulators could leave the market to identify which materials represent an unacceptable problem. This was very much the approach which the industrialized countries have taken throughout their period of industrialization, with new products, materials and processes introduced by their developers as they saw fit. Effectively, the population was treated as the subject in an unofficial, unmonitored experiment; any allocation of responsibility for adverse effects was left to the existing laws on liability. Specific regulation was introduced only at a later stage when any impact on health or the environment had made itself apparent at a sufficient gravity or extent to be noticed. In an age of better educated consumers and a proliferation of consumer interest groups, however, such an approach is no longer politically acceptable.

1.2.6 Political Responses

The recitals to the European Parliament's Resolution of 24 April 2009[9] summarize the problem well:

> F. whereas the current discussion about nanomaterials is characterized by a significant lack of knowledge and information, leading to disagreement starting at the level of definitions:
> (a) concerning the size: approximate indication of the size ('in the order of 100 nm or less') versus a specific size range ('between 1 and 100 nm');
> (b) concerning different/new properties: different/new properties due to size effects, including particle number, surface structure and surface activity, as an independent criterion versus using such properties as an additional criterion for the definition of nanomaterials;
> (c) concerning problematic properties: limitation of the definition of nanomaterials to certain properties (e.g., insoluble or persistent), or not making such limitations.
>
> G. whereas a fully developed set of harmonized definitions is not currently available although a number of international standards are either available or in progress, defining 'nanoscale' as 'having one or more dimensions of the order of 100 nm or less', and often distinguishing between:
> – nano-objects, defined as 'discrete pieces of materials with one, two or three external dimensions at the nanoscale', that is, as materials constituted by isolated objects with very small dimensions;

9. European Parliament resolution of 24 Apr. 2009 on regulatory aspects of nanomaterials (2008/2208(INI)).

- nano-structured materials, defined as materials 'having an internal or surface structure at the nanoscale', for example, exhibiting cavities of small dimensions.

The Parliament therefore called for the introduction of a comprehensive science-based definition of nanomaterials in Community legislation. In response, the European Commission's Joint Research Centre produced a report[10] (the JRC Report) in 2010 but concluded, no doubt to the frustration of Members of the European Parliament, that:

- For a definition aimed for regulatory purposes the term 'nanomaterial' in its current general understanding is not considered appropriate. Instead, the more specific term 'particulate nanomaterial' is suggested.
- The term 'material' is proposed to refer to a single or closely bound ensemble of substances at least one of which is a condensed phase, where the constituents of substances are atoms and molecules.
- For a basic and clear definition of 'particulate nanomaterial', which is broadly applicable and enforceable, it is recommended not to include properties other than size.
- For the size range of the nanoscale, a lower limit of 1 nm and an upper limit of 100 nm or higher should be chosen.
- The questions of size distribution, shape and state of agglomeration or aggregation, may need to be addressed specifically in subsequently developed legislation. It is also likely that certain particulate materials of concern that fall outside a general definition might have to be listed in specific legislation.
- Additional qualifiers, like specific physico-chemical properties or attributes such as 'engineered' or 'manufactured' may be relevant in the scope of specific regulations.

In other words, a specific definition is likely to be needed for each and every area of regulation and these may differ from each other significantly as to upper size limit, particle size distribution, shape and state of agglomeration or aggregation as well as the inclusion of lists of identified materials. Similarly, the 2011 policy principles for the United States recognize the limitations of a purely size-based definition:[11]

> A focus on novel properties and phenomena observed in nanomaterials may ultimately be more useful than a categorical definition based on size alone.

The first definition to be proposed for US regulatory purposes, however, dismisses the combination of size, unique or novel properties compared to larger particles of the same material and being engineered to take advantage of these unique

10. *Considerations on a Definition of Nanomaterial for Regulatory Purposes* at <http://ec.europa.eu/dgs/jrc/downloads/jrc_reference_report_201007_nanomaterials.pdf>.
11. *Policy principles for the U.S. Decision-Making Concerning Regulation and Oversight of Applications of Nanotechnology and Nanomaterials* 9 Jun. 2011 at <www.whitehouse.gov/sites/default/files/omb/inforeg/for-agencies/nanotechnology-regulation-and-oversight-principles.pdf>.

properties as not being readily workable in a regulatory context because of the subjectivity of terms such as 'unique or novel properties' or 'manufactured or engineered to take advantage of these properties'. The definition put forward for comment by the Environment Protection Agency (EPA) in connection with pesticide regulation is therefore based on purely objective criteria, namely size.[12]

An intermediate solution proposed by Maynard[13] is to take a list of 9 or 10 attributes including size and surface area, for each of which a threshold value would be set for each compound, and use these as a collective trigger. This would build on the work already undertaken at the Organisation for Economic Co-operation and Development (OECD) under its Sponsorship Programme[14] to investigate the attributes of a list of 13 key nanomaterials,[15] but assumes a body of scientific knowledge of the properties of all nanomaterials of interest which neither exists at present nor is likely to do so in the near term. As Maynard accepts, much more research would be needed to identify the threshold values, and they would also need to be adjusted as the body of scientific knowledge expands. But developing such a set of information will take a considerable time, and the political and ethical quandary as to what to do about the increasing use of nanomaterials of unknown properties in the marketplace is growing increasingly acute. The responses settled upon by the regulators for different countries and regions are likely to be strongly influenced by the prevailing political culture in each case.

The task of defining nanomaterials in order appropriately to regulate them is thus a serious challenge. Nonetheless, political pressure for some regulation to be applied is leading to the gradual introduction of regulations in several countries and regions. The European Commission consulted in 2010 on a draft Recommendation for a definition for use in future European legislation. The draft definition was:

1. Nanomaterial: means a material that meets at least one of the following criteria:
   - consists of particles, with one or more external dimensions in the size range 1 nm–100 nm for more than 1% of their number size distribution;
   - has internal or surface structures in one or more dimensions in the size range 1 nm–100 nm;
   - has a specific surface area by volume greater than 60 m^2/cm^3, excluding materials consisting of particles with a size lower than 1 nm.
2. Particle: means a minute piece of matter with defined physical boundaries (ISO 146446:2007).

This draft was the subject of intense discussion and lobbying over many months, resulting in substantial changes – and trade-offs – in the final text.

---

12. Federal Register, vol. 76, no. 117 (Friday, 17 Jun. 2011), 35383–35395.
13. *Don't Define Nanomaterials.* Nature 7.7.11, 475 at 31.
14. Discussed at Ch. 5 below.
15. See *List of Manufactured Nanomaterials and List of Endpoints for Phase One of the Sponsorship Programme for the Testing of Manufactured Nanomaterials: Revision* OECD ENV/JM/MONO (2010) 46.

The final Recommendation was released on 18 October 2011.[16] It reads as follows:

> 2. 'Nanomaterial' means a natural, incidental or manufactured material containing particles, in an unbound state or as an aggregate or as an agglomerate and where, for 50% or more of the particles in the number size distribution, one or more external dimensions is in the size range 1 nm–100 nm.
>
> In specific cases and where warranted by concerns for the environment, health, safety or competitiveness the number size distribution threshold of 50% may be replaced by a threshold between 1 and 50%.
>
> 3. By derogation from point 2, fullerenes, graphene flakes and single wall carbon nanotubes with one or more external dimensions below 1 nm should be considered as nanomaterials.
>
> 4. For the purposes of point 2, 'particle', 'agglomerate' and 'aggregate' are defined as follows:
> (a) 'particle' means a minute piece of matter with defined physical boundaries;
> (b) 'agglomerate' means a collection of weakly bound particles or aggregates where the resulting external surface area is similar to the sum of the surface areas of the individual components;
> (c) 'aggregate' means a particle comprising of strongly bound or fused particles.
>
> 5. Where technically feasible and requested in specific legislation, compliance with the definition in point 2 may be determined on the basis of the specific surface area by volume. A material should be considered as falling under the definition in point 2 where the specific surface area by volume of the material is greater than 60 m^2/cm^3. However, a material which, based on its number size distribution, is a nanomaterial should be considered as complying with the definition in point 2 even if the material has a specific surface area lower than 60 m^2/cm^3.

The Recommendation is precisely that – not a binding definition but a proposal to be considered in all future regulations dealing with nanomaterials. Nevertheless, it has been the subject of considerable criticism. The selection of a size limit of 1–100 nm has raised concerns that larger particles which display properties particular to the nanoscale are absolutely excluded. The particle size distribution selected is also contentious, since SCENIHR had recommended a limit of 0.15%, several hundred times smaller than the 50% which appears in the final text without any apparent scientific justification. As a result, where the definition is implemented unchanged, many products containing a large proportion of particles less than 100 nm in dimensions will not be categorized as nanomaterials for European regulatory purposes.

---

16. Recommendation 2011/696/EU (OJ L 275/38 of 20.10.11); full text reproduced at Appendix IV.

Although this limit is subject to variation case by case, the minimum limit of 1% is still 7 times greater than SCENIHR's proposal, and can be substituted only where 'warranted by concerns for the environment, health, safety or competitiveness'. Accordingly even such variation requires some evidence showing particular problems may arise, which appears to take a relatively restrictive attitude to the precautionary principle, given that in default a very high proportion of particles will have to be nanoscaled in order for the definition to capture the material at all.[17] The list of factors which may result in a variation itself demonstrates the political and commercial, rather than purely scientific, nature of the Recommendation, since the inclusion of competitiveness as a factor points to potential future uses of modified forms of the definition to raise barriers to some products independently of their environmental or health impacts. In practice, the form of definition to be used for regulation of any particular sector will now be subject to further lobbying and debate as different interest groups attempt to make the case for a higher or lower threshold within the 1%–50% band.

The Recommendation also proposes that the definition, and in particular the particle size distribution threshold, be reviewed by December 2014 to assess whether any changes are required in the light of experience, scientific and technological developments.

1.2.7 Nanoscale Devices

The regulation of nanotechnology in the sense of devices and structures on the nanoscale, however, does not necessarily raise different issues from either the regulation of macro-scale devices or that of nanoscale materials, until the question of truly self-replicating machines arises. The doomsday scenario of Planet Earth being over-run with self-replicating nanomachines (grey goo), eventually consuming all matter, is still wholly science fiction. Various universities are working on macro-scale self-assembling machines[18] or machines which self-replicate in a limited sense, such as printing out copies of their own components given a suitable feedstock, but as yet no machine which is able to make a direct copy of itself from raw materials without assistance has been developed even on this scale.[19] On the nanoscale, basic functional components such as molecular switches and a 200 nm electric motor[20] have been developed. But nothing on the lines of self-assembling molecular scale machines is likely to be developed in the near or even medium term future.[21] Regulation would be premature, to say the least.

17. Discussed in Ch. 4 below.
18. A machine able to create a copy of itself from a kit of parts – see Zykov et al. *Self-reproducing Machines*. Nature 435 (7038) (2005): 163–164.
19. Jones et al. *RepRap – The Replicating Rapid Prototyper*. Robotica 29 (2011): 177–191.
20. 'World's smallest electric motor made from a single molecule' *Science Daily* 5 Aug. 2011.
21. This raises validity (sufficiency/enablement) issues in respect of those forward looking patents which have already been granted over them, such as US 6,510,359 to Zyvex Corp for a 'Method and system for self-replicating manufacturing stations' (due to expire in 2020, well before any

1.3 HISTORY AND CURRENT COMMERCIAL STATUS

1.3.1 Facts and Stories

There are many potential ways to frame the story of nanotechnology, and it is popular to credit Richard Feynman with the initial idea of manipulating matter from the atomic level upwards as opposed to from the bulk level downwards. Feynman certainly identified that possibility[22] but whether it is fair to say that his idea actually caused any of the later work to be done, or influenced its direction, is debatable[23] since none of the basic tools for manipulating atoms were developed for another two decades. However, the more prosaic alternative history – starting from the development at IBM in 1981 of the scanning tunnelling electron microscope and subsequently of the atomic force microscope, instruments which do allow the nanoscale to be visualized and manipulated – lacks the highly effective association of the new story with a widely recognized brilliant individual.

The reality, of course, is more complex than a single continuous development from just one original insight. For a start, as discussed in the previous section a number of materials have been in use in the form of nanoparticles for many years without being categorized as nanomaterials.

Bearing in mind that nanotechnology's toolkit of techniques and devices is applicable to almost all areas of science, the collection of products and materials which exist or are under development today come from many different scientific disciplines. At the near-atomic level the distinctions between chemistry and physics start to erode, and in practice much work in nanotechnology has involved groups drawing on a wide range of expertise in different fields, rather than 'nanotechnologists' parachuting in to apply what they know to the existing know-how of any given field. The electrical conducting properties of single walled carbon nanotubes (CNTs) were predicted by one group, for instance, before they were actually made by an entirely separate research team.[24]

This tale of diligent laboratory work shared among many people in many disciplines is out of key with the classic narrative of the solitary genius toiling away and single-handedly arriving at a 'eureka' moment after which all difficulties fall away. But it is a more accurate description of modern research, and avoids the risk of the technology becoming associated also with the moral position of a single originator. This is an important issue since, although nanotechnology research produced little with any commercial potential until the late 1990s, the existence of such research reached popular consciousness well before then with the

---

such nano-manufacturing stations can be implemented). See Ch. 2 for discussion of patent law as applied to nanotechnology.

22. *There's Plenty of Room at the Bottom* talk given to the American Physical Society in 1959; first published in Caltech Engineering and Science, vol. 23, no. 5 (February 1960), 22–36.
23. Toumey. *Reading Feynman into Nanotechnology: A Text for a New Science*. Techné 13(3) (2008): 133–168.
24. Iijima et al. *Single-Shell Carbon Nanotubes of 1-nm Diameter*. Nature 363 (6430) (1993): 603–605.

publication in 1986 by Drexler of his book *Engines of Creation* speculating about the future of nanotechnology, both as to its benefits and risks.[25] In particular Drexler raised the spectre of a self-replicating nanomachine out of control and potentially over-running the environment. Self-replicating machines were at the time even further from reality than they are today, but the concept was widely disseminated and led to a degree of public anxiety about nanotechnology similar to that which was current at the time around biotechnology and in particular genetic engineering. This was compounded by a high-profile article in *Wired* in 2000, whose author argued that technologies under development today, including nanotechnology, present a greater threat to humankind as a species than any previous technologies have done, in part due to their ubiquity (compared to previous dangerous technologies such as nuclear power which were available only to limited participants).[26] Nanotechnology became a vehicle for the projection of sometimes nebulous concerns about the future of society in the face of technologies which are increasingly beyond the comprehension of most individuals.

1.3.2 Public Anxiety

The most potent response to such concerns would be to explain exactly what is being developed and its potential uses and risks. Since there was, in fact, little to disclose at the time other than basic research – and in particular no evidence of specific health or environmental risks[27] – researchers and companies active in the field remained relatively silent, which may have compounded the public anxiety by creating an impression of secrecy. Of course, the normal requirement to keep technical developments secret until any application for patent protection has been filed, in order to protect market share once the product has been commercialized, can also feed suspicion since, as discussed below, patenting is itself is sometimes seen as a potentially unjust and/or unethical approach to new technologies.

The debate reached a relatively objective stage with the preparation in 2004 of a report by the Royal Society, the United Kingdom's foremost scientific institution, jointly with the Royal Academy of Engineering, on *Nanoscience and nanotechnologies: opportunities and uncertainties* (the Royal Society/RAE Report).[28] Although possibly initiated as a result of the concerns over self-replicating machines, the report instead focuses on issues surrounding nanomaterials and in particular the issues of potential toxicity associated with nanoforms of known materials. This report has significantly influenced the subsequent legal debate.

---

25. *Engines of Creation: the Coming Era of Nanotechnology* (New York: Anchor Books, 1986).
26. Joy. *Why the Future Doesn't Need Us*. Wired (April 2000).
27. The absence of evidence is not, of course, evidence of absence of risk.
28. Available at <www.nanotec.org.uk/finalReport.htm>.

1.3.3 Commercial Reality

In the meantime, since the first years of the twenty-first century, products incorporating new nanomaterials have begun to be used commercially. The US Project on Emerging Nanotechnologies began in 2005 keeping an inventory on nanotechnology-based consumer products available on the market,[29] at which time only 212 such products were identified on the global market. In March 2011 the figure had reached over 1300 and was continuing to rise steadily. More than half of all products were in the health and fitness category, including clothing, sunscreens, other cosmetics, personal care, filtration such as air and water purifiers and sports goods. However, the inventory is not complete since it is based only on manufacturers voluntarily reporting the use of nanotechnology in their products, which many may be choosing not to do in view of the public anxiety around the term. (Conversely, where a high-technology image may be expected to increase sales, products may be labelled or named to suggest the use of nanotechnology even where this is not the case. There is no legal restriction on the use of pseudo-scientific 'puff' in advertising unless it is actively misleading.) The absence of any complete and reliable inventory or labelling requirement enabling consumers to manage their exposure to nanomaterials is another source of concern to consumer and public health groups.

Given the lack of complete data even as to what products use nanomaterials, let alone what nanomaterials those are, it is clearly not possible to carry out any sort of assessment as to any safety issues which may have arisen. The first safety scare around nanotechnology in Europe concerned a bathroom sealant branded as MAGIC NANO which caused breathing difficulties in some users. However, the product was part of a range all of which contained the same nano-silica particles, yet none of the other products had caused any ill-effects.[30] It therefore appears that the problem was unrelated to the use of the nanomaterial; the fault may have been due to the aerosol propellant. The only other European nano-product recalled to date again concerned an aerosol sealant using fluorsilanes.[31]

Thus, at present a wide range of products either incorporating nanomaterials or manufactured using nanotechnology are available on the market without any safety assessment beyond that stipulated for the bulk form of the substance having been required. While some have been marketed for close to a decade, such that acute problems could be expected to have been identified by now, so little is known about the behaviour of these materials in the human body or the environment that it is possible that problems are accumulating undetected, to emerge only when concentrations reach a threshold level or examples become widespread enough for the causal connection to be recognized.

---

29. <www.nanotechproject.org/inventories/consumer/>.
30. 'Has all the Magic Gone?', *The Economist* (12 Apr. 2006).
31. RAPEX search on 7 Sep. 2011. A third nano-named product, a hair straightener called Nano-keratin, has also been withdrawn but this was due to the fact it contained more than 2.15% formaldehyde; there is nothing to suggest it contained any nanomaterial.

1.4 THE PERCEIVED HAZARDS OF NANOTECHNOLOGY

1.4.1 Introduction

Some of the anxiety surrounding nanomaterials has its roots in a very long-standing mistrust of technology, and substances perceived or classified as artificial, generally. The alchemists of the Middle Ages were viewed with great suspicion by their peers for their attempts to synthesize various substances even where the same substances occur in nature. The divergence between science and other areas of knowledge accelerated in the nineteenth century, and philosophical discussion of the role of technology in society has also tended to a critical approach, focusing on the tendency of technology to prioritize efficiency over other criteria for human life.[32] The perception of a lack of regard for human values and suspicion of motives attaches most easily to technologies which only highly specialized scientists and technologists have any real claim to understand, as is the case with most of nanotechnology.

This section therefore looks at what little is actually known so far about the risks posed by nanomaterials.

1.4.2 Complexity of the Problem

The whole potential of nanotechnology lies in its ability, through the manipulation of materials and devices at the very small scale, to produce effects which are not produced by existing materials and devices on the micro- or larger scales. Although particles at nanoscale both occur in nature and have been produced by industrial processes for many years, these were undetectable and their properties, quantity and distribution were all unknown. They do not provide useful experience which could be used to forecast the impact of the new generation of products and materials.

Research into the properties exhibited by materials on the nanoscale has been going on for some years, but the information available is far from complete. The OECD Sponsorship Programme will, when it reports in 2011, provide the first complete characterization of a set of commonly used nanomaterials, which will provide the basis for comparative work in the future. For the moment, however, it will establish the properties of only the 13 materials which have been under investigation – a tiny fraction of the total number of materials which are in use or in development.

The novel properties of nanoscale materials (compared to the same chemical substances at larger scale) include surface effects such as electrical charge, ability to adsorb onto other particles, tendency to aggregate and agglomerate, crystal structure and stability. These may vary between different forms of the same

32. For instance Ellul, *The Technological Society*. Trans. Wilkinson (New York: Alfred A. Knopf, 1964).

material, for instance where a nanoparticle is coated or uncoated. Once nanomaterials are combined together to create nanoscale structures, the range of possible unfamiliar characteristics escalates rapidly.

It may well be that many such properties will have either no discernible effect on human health or the environment, or no adverse effect. But in the present state of knowledge, it is impossible to say which properties may cause what intentional or unintentional effects when the material or device is outside the controlled manufacturing environment, or released from the macro-scale product into which it has been fabricated. The issues must therefore be considered at all stages in the material's potential lifecycle since a material which is harmless when incorporated into a solid matrix may – indeed is likely to – have entirely different properties when collecting in groundwater or airborne. These different properties may or may not be harmful, but they make it essential to consider the material or product in each context separately rather than applying a 'one situation accounts for all' approach to evaluating the potential hazard. Of course, macro-scale materials also can and do display different properties when incorporated in a solid compared to in suspension or solution in liquid, or entering the body through different routes – water is toxic if inhaled, for instance. But the relative familiarity of most materials on the macro-scale means that manufacturers and consumers are both confident in handling them appropriately: there is no need for a regulation to prohibit or protect people from breathing in water.

1.4.3 Causes for Concern

The one thing that is known about the whole category of materials on the nanoscale is their size, placing them when free in air into the known hazard category of fine dusts. Fine dusts are hazardous for several reasons, including the tendency to cause explosions in enclosed spaces, and less dramatically because of their ability to enter the body through a number of different routes including inhalation, the eye, but also ingestion (where they may have the ability to permeate the intestinal lining where larger particles do not) and even through the skin, particularly when damaged by injury or disease but also potentially through normal conduits such as the pores. The mechanism by which fine fibres of asbestos – in the region of 60–100 nm long – cause disease are still not fully understood, but it seems that it is mechanical rather than chemical (i.e., results from their size, shape and physical properties rather than their chemical nature).[33] There are basically three hypotheses regarding how asbestos brings about diffuse malignant mesothelioma (a form of cancer rare save in association with exposure to asbestos). The first is the 'oxidative stress theory' is based on the fact that phagocytic cells that engulf asbestos fibres produce large amounts of free radicals due to their inability to digest the fibres, and epidemiological studies indicating that iron-containing asbestos fibres appear

33. Manning et al. *Diseases Caused by Asbestos: Mechanisms of Injury and Disease Development.* Int. Immunopharmacol 2 (2–3) (February 2002): 191–200.

more carcinogenic. Second, the 'chromosome tangling theory' postulates that asbestos fibres damage chromosomes when cells reproduce. Third, the 'theory of adsorption of many specific proteins as well as carcinogenic molecules' states that asbestos fibres in the body concentrate proteins or chemicals including the components of cigarette smoke.[34] The same mechanisms could in principle apply to cause disease when the fibres entering the body share the same physical dimensions, as high aspect ratio carbon nanotubes can do, even though chemically distinct.

The example of asbestos is particularly pertinent when considering the concerns facing regulators around nanotechnology. The increasing use of asbestos in construction from 1880 to 1930 led to increasing exposure both of workers and consumers. However, even though suspicions as to the health impact of asbestos on workers began to be raised early in the twentieth century, it took until 1931 for the first Asbestos Industry Regulations to be introduced, in the United Kingdom. By the time in the 1970s when major health concerns about asbestos had led to strict restrictions on its use, asbestos – cement products were ubiquitous in numerous countries around the world. Bans were not introduced until the 1980s, and some uses remain legal.

One characteristic which so delayed the introduction of regulation was the fact that asbestosis might show only limited symptoms while the exposure was ongoing, but is able to cause death up to 60 years after exposure, with a death toll in the United Kingdom expected to continue to rise, despite the massive reduction in exposure, until 2015.[35] The fact that an individual cannot even detect at the time of exposure that fatal damage is being done makes the threat, once realized, more frightening than more overt dangers.

This fear of being harmed by an unknown, unseen agent against which the victim is unable to defend herself at the time, transposes easily from the case of asbestos to the case of engineered nanomaterials. This awareness has triggered some of the interest in the potential health effects of nanomaterials, and carbon nanotubes and other high aspect ratio nanoparticles in particular, given their similarity to asbestos fibres in terms of size and shape. Early research seems to suggest that inflammation of the lungs and other conditions reminiscent of asbestosis may result from inhalation of carbon nanotubes.[36] The extent of the effect depended, among other things, upon the precise length of the nanotubes.

A review of the studies available in 2008 as to the potential toxicity of other nanoscaled materials already in large scale use, such as fumed silica, titanium dioxide and the nanoscale encapsulation systems used for various food and cosmetic applications, concluded, despite finding no accepted reports of any

34. Toyokuni. *Mechanisms of Asbestos-Induced Carcinogenesis*. Nagoya J Med Sci. 71(1–2) (February 2009): 1–10.
35. Final Regulatory Impact Assessment for the Control of Asbestos Regulations 2007, *Health and Safety Executive* 2007.
36. Poland et al. *Carbon Nanotubes Introduced into the Abdominal Cavity of Mice Show Asbestos-Like Pathogenicity in a Pilot Study*. Nature Nanotechnology 3 (2008): 423.

toxicity among the substances studied, that a case-by-case approach was, nevertheless, necessary because of the absence of generalized observations that would apply to all nanosized materials.[37] As if to prove the point, the European Food Safety Agency (EFSA) released an opinion later the same year in which it concluded it was not possible on the data available to assess the safety of nanosilver as a food additive.[38] Minimum requirements include data from repeated-dose toxicity studies and appropriate *in vitro* studies (e.g., for genotoxicity). The design and performance of these toxicity studies must be based on toxicokinetic information on the nanomaterial. However, the dossier provided for the purpose of the opinion inadequately characterized the silver hydrosol in question and no evidence was given for the form of silver to which cells or animals were exposed in those tests which had been carried out, although the petitioner provided additional information on the toxicity and bioavailability of ionic silver.

A 2010 review[39] concluded that:

> Many studies have examined the pro-inflammatory effects of manufactured nanoparticles, on the basis that their ability to cause inflammation is a major predictor of potential hazard in such particles. The first important finding was that nanoparticles have a more pronounced effect on inflammation, cell damage and cell stimulation than an equal mass of particles of the same material of greater size. This appears to hold true for materials as varied as carbon black, titanium dioxide, various metals and polystyrene. Surface area is the metric driving the pro-inflammatory effects and this is evident both *in vitro* and *in vivo* particles of various sizes producing inflammatory effects that are directly related to the surface area dose.

What this conclusion does not say, is that the materials studied have in fact been shown to cause disease or death when incorporated in products or in an amount or context likely to be encountered in the normal course of use. It concludes no more nor less than that gram for gram, the nanoforms of substances are more likely to cause harm than the bulk forms. But the amounts which would be necessary to cause any particular harm through any particular exposure have not been identified, and nor has the likelihood of exposure to such amounts for anyone other than a worker whose occupation involved handling nanoparticles been assessed.

Nanomaterials have also been reported to have health impacts on fish, insects and other non-mammalian species,[40] which could lead to these materials being

---

37. International Risk Governance Council Report, *Risk Governance of Nanotechnology Applications in Food and Cosmetics*, 2008.
38. *Scientific Statement of the Panel on Food Additives and Nutrient Sources added to Food on a* request *from the Commission on Silver Hydrosol as a Source of Silver Added for Nutritional Purposes to Food Supplements.* The EFSA Journal (2008): 884, 1–3.
39. Seaton et al. *Nanoparticles, Human Health Hazard and Regulation.* J.R. Soc. Interface 7 (2010): S119–S129. References in the original omitted.
40. Handy et al. *The Ecotoxicology and Chemistry of Manufactured Nanoparticles.* Ecotoxicology 17 (2008): 287–314; Klaine et al. *Nanomaterials in the Environment: Behavior, Fate, Bioavailability, and Effects.* Environ. Toxicol. Chem. 27 (2008): 1825–1851.

classified as harmful or toxic to non-mammalian species.[41] But no such classification has yet taken place.

Where the nanomaterial itself is not toxic or liable to trigger immune system responses, it may still act as a Trojan horse, providing a conduit for other toxic materials to enter the body or within the body, to enter specific organs or cells.[42] Hollow carbon nanotubes are being considered for use as a vehicle for drug delivery within the body, but the same structure may cause carbon nanotubes intended for other applications to take up other materials and aid these to enter the same cells. This does not mean, however, that all nanoscale materials will possess the particular binding qualities to enable this to happen, rather, once again it will need to be assessed case by case – unless and until the characteristics which enable such binding are sufficiently well understood to make it possible to predict with confidence which groups of nanomaterials will share that property.

1.4.4 Exposure Paths

It can be assumed that deliberate manufacture will increase the quantity of engineered nanomaterial to which an average individual will be exposed (whether as worker or user/consumer), and it is likely that as a result the range of nano-specific properties the individual is likely to encounter will also expand. Exposure to humans can occur at any stage in the lifecycle of a nanomaterial: manufacturing; application into a finished product; behaviour when the product is in use (do the nanoparticles remain embedded, wash off, permeate the user's skin?); or after disposal when present in the environment either in soil, water or air. The material may be presented to the body in different forms and through different routes at these different stages. For instance, one established product using nanoparticles (of titanium dioxide) is sunscreen. Initially free particles are incorporated into the cream during manufacture. The consumer then applies the bulk of the cream to the skin, but it may wash off due to sweating, swimming or deliberate washing, ending up in waste water or swimming facilities. The remainder in the 'empty' tube is disposed off as household waste and may be incinerated, releasing the particle into the air, or deposited in landfill, where the particles may become incorporated with other materials or may leach into ground water. From waste or ground water, it may pass through a purification process and be returned to the consumer in the drinking water supply.

Once released into the environment, the particles may aggregate, adsorb onto other particles (of soil, or other man-made materials) or remain as distinct particles. In any of these forms they may affect different components of the eco-system,

41. Kahru and Dubourguier. *Review: From Ecotoxicology to Nanoecotoxicology*. Toxicology 269 (2010): 105–119.
42. Ashwood, et al., *Fine Particles That Adsorb Lipopolysaccharides via Bridging Calcium Cations May Mimic Bacterial Pathenogenicity towards Cells*. Exp Biol Med 232(1) (2007): 107–117.

being taken up by plants along with water, absorbed by nematodes and amphibians through their skin, or swallowed by fish and animals breathing or drinking.

The concentration of nanomaterials in the environment will depend on factors such as the nature and amount of the nanomaterial released, its physical and chemical properties, and time. Nanomaterials released into the environment may undergo transformation by environmental conditions such as temperature and salinity, biological conditions such as habitat, and the presence of co-contaminants. In turn, the transformed nanomaterials may modify atmospheric, soil, or water chemistry. These transformations may alter the form of the nanomaterials to which humans and ecosystems are exposed and which are transported through the environment. Biological or environmental systems may be exposed to these dispersed engineered nanomaterials and respond through systems and pathways designed to buffer exposures to substances that could perturb human health or adversely impact the environment.

As a result, the potential sites at which exposure to humans could occur and the form in which the nanomaterials to which they are exposed, are as diverse as the range of contexts in which the particles' behaviour needs to be understood. And different nanoparticles may behave differently in each of these contexts. What behaviours actually occur is unlikely to become apparent until particular kinds of nanoparticle have been in use in sufficient quantities for a sufficient length of time, for those properties to be observed. It therefore simply cannot yet be said what probability of harm arises through any given route. As was the case in respect of asbestos and, indeed, lead in air as a result of petrol combustion, it may then take some time before any connection is established between any observed health effects and the underlying cause.

Nevertheless, the perceived risks arising from nanotechnology escalate as the volume of nanomaterials produced and circulated to consumers increases: even if the actual risk from exposure remains stable (or unknown), the increasing extent of exposure increases the likelihood of cumulative harm. The question of how to address this perceived risk is one of the most challenging for regulators.

1.5 NANOTECHNOLOGY AND ETHICS

1.5.1 Introduction

The combination of widespread anxiety and very high expectations as to the potential benefits of nanotechnology has led, as it did with biotechnology, to major ethical questions being raised as to the purpose for which nanotechnology should be regulated. Opponents of biotechnology feared the potential for contamination of natural organisms by transfected genes and the risk that the cost of new biotechnology-enabled medicines would mean that the poor disproportionately suffered from diseases of which the rich could be cured. Opponents of nanotechnology fear long-term exposure to nanomaterials may bring unforeseen public health risks, and that environmental contamination may cause damage to critical ecosystems, which may be irreversible by the time it becomes apparent. Concern over the safety of nanotechnology has reached such heights in some quarters that a number of organizations suggested to the European Commission's scientific hearing on risk in nanotechnology the imposition of a moratorium on sale of foods or food packaging incorporating nanotechnology pending research demonstrating its safety. While ethically unfaultable from the point of view of risk-avoidance, however, such an approach also risks making the development of useful technologies too slow and expensive to be done on a commercial basis since foods and food packaging will never command the premium prices which enable pharmaceuticals to be developed despite such stringent safety requirements. Such research is very unlikely to be carried out through public funding, and therefore may not be done at all. The ethical balance between the security of current food products and risk of foregoing the potential (but unproven) advantages of new packaging and additives is not an easy one to strike. The balance may, indeed, be in favour of foregoing further research except where the benefit of improved food packaging is such as to rival the potentially life-saving benefit of new pharmaceuticals. But in developed countries food security is already sufficient (if not perfect) without the possible enhancements of nanotechnology; if such an ethical position were to be taken, these areas of research could have to await philanthropic or public funding aimed at alleviating problems of food spoilage in developing countries.

Management of such uncertain risks is made more difficult by the lack of agreement on what risks might be socially acceptable or unacceptable. The answers to such questions are likely to depend on local circumstances and culture. The risk of an unknown impact on health from the use of nanotechnology in food packaging is likely to have a completely different importance to a starving man than it does to a well-nourished one. The ethics of nanotechnology may need to focus on different issues when considered from a global, regional or national perspective.

1.5.2 Moral Issues

Complex though environmental issues are, the moral issues arising from nanotechnology are more complex still given that they include not only the fundamental question of humanity's effect on the natural systems which support it and the possible health implications of exposure to different nanomaterials, but also humanity's relationship with itself, through the possibilities for enhancement of the body offered by developments in nanomedicine. Whereas medicine in the sense of repair or healing of disease and injury is familiar ethical territory, the potential to go routinely beyond the body's natural abilities requires an answer to the question of what it is to be human – an answer which must surely include the ability to recognize and grapple with moral and ethical as well as practical issues. The moral issue here is whether, by treating the body as a mechanical, biochemical system which can be tinkered with in the same way as an industrial process or advanced engine system, society risks reducing the value of individuals as human beings by reference to the perfection (or otherwise) of the body they can afford. The imbalance entrenched by such a reductive approach is already evident in the consumer society as it exists today, with ever younger people undergoing cosmetic surgery. The consequence of reinforcing that trend to include enhancements to physical performance and function should be considered with great care: is the best course simply to make it possible and deal with the effects once they have been realized?

It is, however, unlikely given the established decision-making apparatus in developed countries and the difficulty the international community has had in formulating a coordinated legal approach in any field except the laws of war and trade, that there will be any international consensus or state intervention over such applications. Certainly where issues of choice and morality are concerned, the range of cultures and values participating in international discussions renders such questions imponderable. Nor do the major industrial nations currently have regulatory bodies competent in moral or ethical matters – indeed, it is one of the hallmarks of modern democracies that subject to the limits of the criminal law citizens are free to hold whatever beliefs and practise such modes of life they see fit. Davies' proposal for new forms of oversight which would take into account the challenges raised by nanotechnology to moral and ethical beliefs and social impacts which are not currently dealt with by regulatory frameworks[43] is unlikely to be feasible within any nation other than the smallest and most culturally uniform – certainly not within the United States or European Union (EU). In this connection it is notable that one of the smallest and most culturally uniform of international organizations – the OECD – has so far been the most active international forum for coordination efforts by regulators and industry experts. But it is quite likely that it has achieved the successes it has precisely because it is a relatively exclusive club focused on economic issues. Its work is unlikely to give rise to any global consensus on a broader approach to regulation.

In any event, as the authors of the 2006 United Nations Education, Science and Culture Organization (UNESCO) Report *The Ethics and Politics of Nanotechnology*[44] (the UNESCO Report) point out, the technological repair, replacement or enhancement of human beings using sensors, implants and so forth to create a category of 'post-humans' is one which arguably exists already independently of nanotechnology and its present or future developments. Performance enhancing drugs are used in sport, retinal implants enable the blind to experience partial sight and hearing implants are almost routine enhancements, albeit in the context of those whose hearing is otherwise diminishing. The ethical issues as to such existing technological interventions into the human body are qualitatively similar to those potentially raised by future nanotechnological advances. The UNESCO Report therefore categorizes this supposed issue as a distraction so far as nanotechnology governance is concerned, liable to divert precious attention away from the genuine

43. J. Clarence Davies, *Oversight of Next Generation Nanotechnology*, Project on Emerging Technologies PEN 18 (2009).
44. Available at <http://unesdoc.unesco.org/images/0014/001459/145951e.pdf>.

nano-specific concerns to which governments and non-governmental organizations could more profitably address themselves.[45]

1.5.3 Socio-Economic Issues

1.5.3.1 North–South Equity

Another ethical argument raised is that the breakthrough new products which are being developed must be managed in the interests of all rather than entrenching further the socio-economic inequalities of developed versus developing nations. The potential applications of nanotechnology in areas such as water purification in particular are seen as objects which it would be unethical to deny to developing countries, whose populations desperately need such solutions but which are unable either to invest in developing them for themselves or to afford to buy in foreign-owned rights and products (which global patenting may mean are restricted even in countries where the technology is not being exploited by the patent owners).

A further concern is the potentially disruptive impact of nanotechnology on the economies of developing countries, for the majority of which commodity production is the backbone of the economy. Historically, advances in science and technology have had profound impacts on commodity production and trade and there is every likelihood that nanotechnology will in its turn change the commodity markets, disrupt trade and eliminate low-skilled jobs.

Since such issues are intrinsically global, it is not surprising that UNESCO, which has since the 1990s taken an active role in promulgating Declarations intended to establish normative standards in the fields of bioethics, environmental and technology ethics, is among the institutions which have already considered the ethical implications. As the UNESCO Report explains, the extent to which development draws upon the resources of the developing world but primarily benefits the more developed countries is key. Nanotechnology, intersecting with many of the most important fields of development – medical devices, drug delivery, information science and ecology – faces the same ethical issues.

The UNESCO Report acknowledges that whereas the ethical issue of disparity of access as regards biotechnology research was a disparity *between* nations, with the well-documented 'North–South' divide potentially disadvantaging citizens of less developed nations, the internationalization of research and dissemination of information via the internet means that so far as nanotechnology is concerned an equivalent or even greater disparity of access may be *within* nations, between the educated and wealthy elite and the poorer and less well-educated strata of any given society. The ethical question this raises is as to the appropriate direction of

45. In contrast, the European Commission's Action Plan for Europe 2005–9 on nanosciences and nanotechnologies (COM(2005) 243 Final) identified non-therapeutic human enhancement as one of two ethical issues raised by research into nanotechnologies, the other being the invasion of privacy due to invisible sensors.

publicly funded research, which could focus more on the less commercial applications of nanotechnology such as energy storage, water treatment and diagnosis and treatment of medical conditions and diseases common to less developed countries. This of course is the same issue as has already been recognized in connection with pharmaceutical research generally, where some public institutions and major charities have stepped in to try to address the gap left by commercial researchers whose focus is on products which are attractive to the richer consumers of the north and west. As regards the remaining 'North–South' disparity, an ethical position may require capacity building to empower developing countries' researchers to participate in the development of nano-enabled products relevant to their local populations' needs.

1.5.3.2 Monopoly Rights

Concerns are also raised over the extent to which exclusive rights have been registered over the very early developments in nanotechnology, potentially restricting access to the basic technology toolkit for researchers across a wide range of applications.[46] While the ability to patent new developments has applied to all areas of technology for many years,[47] the context as regards nanotechnology is different because attitudes to patenting have fundamentally changed in the last few decades. In the 1960s and 1970s, when semiconductor devices and early biotechnology developments were at the cutting edge of research, universities in particular subscribed to the principle of public ownership of research results, and even IBM opted to disclose its discoveries to the public in a Technical Disclosure Bulletin which represented the gold standard of innovation for much electronics work. However, in 1980 the Bayh-Dole Act was passed in the United States giving universities the right to profit from patenting the results of publicly funded research and in the following decade British universities likewise shifted towards seeing technology innovation as a source of additional income in the face of diminishing government funding. Consequently, nanotechnology is arguably the first major area of research to have been treated as a patent 'land rush' from its very earliest stages.

This concern must, however, be considered in perspective. First, patent rights do not last more than 20 years save for certain US patents[48] applied for before June 1995. Accordingly, for most of the world any patent applied for in or before 1990 had expired by 2010. The technical details of any inventions underlying products launched in the first few years of the twenty-first century will have been patented (if at all) in the late 1990s; such patents will expire in the next few years. Their ability to limit research is in any case constrained, outside the United States, by the

46. See for instance, ETC Group Report *Nanotech's 'Second Nature' Patents: Implications for the Global south* Communiqués no. 87-8, March–June 2005.
47. Subject to local and regional exceptions in areas such as methods of medical treatment.
48. So called 'submarine' patents, whose date of expiry depends not on their date of filing but their date of grant, which may be any number of years later. However, this indefinite duration applies only within the United States and not to any equivalent patents granted in the rest of the world.

existence in many countries' patent laws of an exemption for experimental use which means that research on patented subject matter can be conducted prior to patent expiry even though the results may not be commercialized until afterwards.[49] Thus, the portrayal of the patent landscape in nanotechnology as a long-term absolute roadblock to development of new nanotechnology products in developing countries suited to local needs is largely inaccurate.

Another ethical issue which is raised, but is unfounded, is the argument that patents over nanoscale forms of existing substances are being used to prevent the use of these substances – whether chemical elements or compounds derived from plants and animals – in traditional medicine.[50] One example cited is of patent claims to a nanoscale emulsion containing ginseng for use in cosmetics.[51] However, this and related arguments fundamentally misunderstand the effect of a patent grant, which is limited to the scope of what has been invented. If there is evidence that part of what is claimed has been previously known or used in public, then the claim is invalid to that extent. Accordingly, claims such as the example above are valid only in respect of use of the nanoform; the use of a macroscale emulsion, or any other form of ginseng, as was presumably the traditional usage, is entirely unaffected.[52]

The ethical arguments are not novel, having long been addressed to the issue of pharmaceutical and biotechnology research. The inequity of the use by developed countries' companies of biological material obtained from the flora and fauna of less developed countries to create new medicines unavailable to the citizens of the countries of origin of the materials, without compensation, has been the driving force behind a long-standing negotiation at the World Intellectual Property Organization (WIPO) under the Convention on Biodiversity to require disclosure and acknowledgement of country of origin in any new patent application and proposals have been discussed as to how to share the benefits. The practical difficulties around this – such as the fact that many companies work from materials collected decades ago, which may not be classified as to their exact origin for instance – do not affect the underlying ethical principle, that all parties who have contributed to a new treatment should share in some way in the benefits. In 2001, the World Trade Organization (WTO), a body not previously noted for its social or ethical agenda, declared that member countries should 'promote access to medicines for all'. In 2007, the General Assembly of WIPO adopted 45 recommendations establishing a development agenda, recognizing for the first time that intellectual property (IP) is not a moral abstract but a policy instrument.[53]

---

49. While US law does include an experimental use exemption, in practice it has been interpreted extremely narrowly by the courts such that no research for commercial exploitation is likely to be exempt (*Roche Products v. Bolar Pharmaceutical*, 733 F.2d 858 (Fed. Cir. 1984)).
50. ETC Group Report, *supra* n. 44.
51. Pacific Corp's EP 1327434 *Nanoemulsion Comprising Metabolites of Ginseng Saponin and a Skin-Care Composition for Anti-Aging Containing the Same*.
52. The legal issues around patenting nanotechnology will be discussed in more detail in Ch. 2 below.
53. See <www.wipo.int/ip-development/en/agenda/recommendations.html#c>.

The nanotechnology equivalent argument is, nevertheless, slightly less clear. Development of a new nanotechnology solution to water purification does not depend upon collecting polluted water from a given source, nor does it normally involve the adaptation of traditional knowledge. The only remaining ethical lever is the overall imbalance of wealth, research facilities and incentives between the developed and developing countries. Insofar as much nanotechnology research to date has been carried out within publicly funded institutions such as universities, then clearly it is open to the funding authorities to apply conditions (if government policy so dictates) requiring access to be given to institutions or individuals in developing countries, whether cost-free or on a favourable royalty-bearing basis. The UNESCO Report proposes that all publicly funded research results should be made open access, and preferably in both electronic and print form since many scientists still work in countries where internet access may be either limited or unreliable. This is, however, as much a political as an ethical issue: in financially difficult times, citizens of developed countries may be less sympathetic to their governments' use of tax revenues for what are perceived as aid purposes than they were during the prosperous years.[54]

1.5.3.3 Outsourcing Risk

There are also potential issues over the outsourcing of hazardous manufacturing processes to poorer countries where occupational health and safety standards, even for known hazardous chemicals, are often lower and less effectively enforced than in developing countries. Further, a prime benefit asserted for the new technologies is the reduction in quantity of raw materials which they may require for manufacturing goods – a development which may disrupt the markets for commodities, which are the mainstay of the economies of many developing countries. A new field of manufacturing almost inevitably also implies a new source of waste for disposal, another area in which the developing countries have historically taken a share arguably disproportionate to the benefits. Further, rather than the benefits of nanotechnology solutions resolving existing problems of development, promotion and debate about nanotechnology in developing countries threatens to divert and detract resources, political will and attention from the immediate needs of the poor and could inhibit research necessary to address society's problems in a systemic manner, institutionalizing inequality in this field as in others.[55]

The UNESCO Report addresses the ethical problem of disparity of risk. Clearly, once risks are identified and in particular any associated with the manufacture of nano-enabled technologies, there is an ethical question as to who should bear such risks, how it should be distributed geographically and who will have the power to decide. The fact that nanoparticles released into the atmosphere have the potential

54. 'Soft Power, Hard Hearts', *Economist* 23 Jul. 2011, 25.
55. *Nanotechnology in a Developing Country – Applications and Challenges*, presentation by Prof. Babajide Alo, Ph.D. University of Lagos, Nigeria at IFCS Forum 6, available at <www.who.int/ifcs/documents/forums/forum6/ppt_nano_alo.pdf>.

due purely to their size to be transported all around the globe means this risk may not simply come down to a decision as to where to site manufacturing; the hazard may be distributed independently of the intent of the producer. Likewise, given the dearth of knowledge as to the behaviour of different engineered nanomaterials in the terrestrial and aquatic environments, the problems which may (although it cannot yet be said *will*) arise may have cross-border implications which have been relatively limited so far as existing technologies are concerned. Although, of course, the pollution of oceans and river systems has already become an international issue, it has tended to be addressed (if at all) through bilateral discussions which may simply not be adequate for more dispersed problems.

This pre-supposes of course that the product, including whatever nanotechnology it may, is produced and distributed in a risk-bearing form. The 2006 UNESCO report notes existing research addressing the amelioration of risk relating to fullerenes by the conjugation of the particles themselves to other chemical. It recommends that if such approaches may exist to reduce or even remove any risk associated with the fundamental form of the particles, then ethical considerations should drive the producers towards investigating their availability rather than carrying on to develop marketable products with the 'native' form of the nanoparticle, risks and all. The Royal Society/RAE Report recommended the similar approach of a 2–5-year window for toxicity investigation and management before new regulation should be introduced. However, there is no guarantee that all companies operating in the field are using the present regulatory inter-regnum for that purpose. Indeed, current regulatory control of known highly hazardous materials such as arsenic and mercury is weak in some countries, in which case imposing stricter controls in respect of engineered nanomaterials is politically challenging.

Further, arguments such as these apply equally to other forms of innovation and, were any general answers to be proposed to the questions raised, such answers cannot really be limited to nanotechnologies. At bottom, the issue is the historical and economic relationship between the industrialized and less developed countries and the extent to which moral and ethical standards should influence future political and economic actions. Nevertheless, the debate has arisen in the context of the development of nanotechnology and in view of the relatively less entrenched *status quo* of investment and ownership, proponents of an ethical approach to nanotechnology may hope to begin a shift in attitude which could over time extend to more established fields as well.

1.5.3.4 Transparency

The importance of communication of risk to the public to the same extent as communication of benefits is also under discussion as an ethical principle, since the possibility of exposing the public to hazards of which they are not informed is hardly ethically defensible unless there were some overwhelming public good reason for it. The difficulty here is that the hazards are in the main unknown: while the toxicity of nanosilver, titanium dioxide and carbon nanotubes has been identified, the potential impact of other nanomaterials either on human health or the environment is not yet known. But it is not possible to say there will be none. Such an unclear picture cannot realistically be communicated to any useful effect, particularly given the relatively limited understanding of science among the general public even of industrialized countries. But the public does have a right to access such information as is reliable and to know how to do so; it may be that this is the best that can be done until more of the uncertainties have been eliminated.

The UNESCO Report is ambivalent on addressing consumer awareness and safety through labelling and transparency requirements, since labelling in practice already leads to conflicting and confusing warnings, the augmentation of which is unlikely to increase consumer understanding. Again, better understanding of the potential and actual risks is needed before any consensus would be possible on how such information should be communicated. The report is clear, however, that non-communication, which produced such suspicion and backlash in the case of products incorporating genetically modified components, is unlikely to be a successful strategy regardless of the ethics.

Advocacy groups have also suggested that there should be more transparency around ongoing research, possibly to the extent that the decision as to what directions research should take could be debated before the investment was made. This would, however, need to be managed to avoid conflict with the requirement for secrecy to enable researchers to obtain patent protection – which is normally essential for commercial research funding. A publicly accepted research programme which cannot to be carried out for lack of funding is not obviously a better solution than a funded programme of which the public become aware only through publication and patenting of results after the event. Once again, this issue is not limited to nanotechnology research: the underlying argument that the public should have more engagement in directing research applies to any area of technology and should not be addressed as though it were nanotechnology-specific. It has become part of the debate because of the very high expectations around nanotechnology, possibly a consequence of advocates over-emphasizing the potential benefits in order to allay the potentially equally over-stated concerns reviewed above. But exchanges of exaggerated benefits and risks do not make for a rational basis for decision-making over investment, whether by governments or commercial interests.

Other suggestions for addressing the ethical issues raised by nanotechnology include the regulation of new technologies by a UN body, potentially taking into account the social utility of any particular proposed research project. This would, however, require a fundamental re-structuring of civil society at the global level, and is unlikely to see the light of day. Indeed, the UNESCO Report considers the opposite extreme in the context of the potential for use of nanotechnology in creating new forms of weaponry and the risk of suppression of scientific results as a result – which may be either because there is real potential for such use, or using the risk as an excuse for the suppression of results incompatible with the political agenda of governments where the research is conducted. Organizations such as UNESCO may, however, in the long term be able to facilitate the definition of norms of scientific conduct around these conflicting pressures.

Chapter 2
Exploitation of Nanotechnology

2.1 INTRODUCTION

This chapter looks at a range of issues relating to the legal framework and practicalities for effective commercial exploitation of nanotechnologies, other than the regulatory requirements for particular applications and sectors which are dealt with in Chapter 6. Issues of potential liability under existing laws are covered in Chapter 3.

Intensive investment in research is necessary to bring about new discoveries on the nanoscale and to develop commercial applications for them. Such investment, unless publicly funded, would not be viable unless an adequate return can be obtained through sales of the resulting products. In practice so far as commercial research is concerned, and in particular where third party funding needs to be obtained for the development of a product, this requires the ideas and techniques concerned to be protected against competition. This is the role of IP and trade secrecy. IP comes in a number of different forms, of which the most relevant to nanotechnology are patents and, where they exist, petty patents (otherwise known as utility models).

A patent is a legal right which confers on the holder the ability to prevent any other person from exploiting the patented invention without their permission, for a period of 20 years.[1] Petty patents or utility models confer similar rights, but are granted for less inventive work – lesser improvements to an existing technology for instance – and have a shorter duration of 6- or 10 years.[2]

---

1. Article 33 World Trade Organization Agreement on Trade-related Aspects of Intellectual Property. Extensions are available in some countries for pharmaceutical inventions, for variable periods up to five years.
2. Petty patents are national legal rights and the duration, where they exist at all, is set by each country. Proposals to introduce a harmonized regime of utility models for the EU were dropped after consultation showed that industry was not in favour, in the late 1990s.

All developed countries, and almost all less developed countries, have a patent system since this is required in order to attain membership of the World Trade Organization (WTO). There is, however, no such overriding obligation to have a system of petty patents; they are accordingly available in far fewer countries.

The rights of a patent holder are powerful and therefore valuable, but in return for the grant of a patent the inventor is required to disclose the invention to the public in detail so that, once the patent monopoly expires, the rest of the world can freely exploit it. For most inventions, either the 20-year period covers the maximum predicted lifespan of the product, or patenting is the only viable form of protection since as soon as the product is made public it would be apparent to competitors how to produce their own versions of it, leaving the originator exposed to price competition from a very early stage. But in some cases, where the invention cannot readily be reverse engineered from the product itself, it may be possible to maintain a monopoly in perpetuity by keeping the invention confidential. Since compositions and structures on the nanoscale are still not easily capable of analysis, and some aspects may not be capable of precise analysis at all, trade secrecy may be a viable alternative to patent protection, particularly where manufacturing techniques are concerned. The law on trade secrecy is therefore discussed in the second section below.

The diverse range of technical fields in which nanotechnology may be used means there is more than the usual scope for confusion through the use of different terminologies for very similar concepts. There is therefore considerable effort being put into developing international standards, which include standards for terminology as well as measurement, characterization of properties and health and safety issues. These in turn support the wider exploitation of technology through enabling interoperability, quality assurance and reliable information which can be used to substantiate the benefits and allay concerns which inevitably arise when a new field of technology incomprehensible to the person in the street begins to penetrate the marketplace. Discussion of standards activities constitutes the third section of this chapter.

The same diversity of potential applications for many forms of nanotechnology means that for many new nanoscale inventions the inventor is unlikely to have the range of competencies necessary to develop the invention across its full potential range. A new material may have the potential to be used both in medical devices and also space engineering, to take an extreme example. This means that licensing will often be the most efficient way to maximize the exploitation of an invention. Licensing out rights enables a right holder to select appropriate partners for developing different potential applications, or in different geographical markets, while retaining ultimate ownership and control of the technology; and other rights may need to be licensed in, to exploit the invention in the chosen application. In both cases, the form of agreement will be critical to ensuring a successful relationship. The permutations of possible forms of licence are an endless topic, so the fourth section of this chapter can give only a brief overview of the principal issues.

Commercialization in the form of consumer products inevitably means advertising, which has been one of the topics causing concern among consumer groups.

Although there are no nanotechnology-specific regulations yet in place, the general law on misleading advertising applies and there are calls for these to be reviewed to address nanomaterials in particular. The fifth section below reviews advertising law in the EU and the issues which are being raised.

2.2 PATENT ISSUES

2.2.1 THE NATURE OF PATENT RIGHTS

The monopoly given by a patent is not a positive right to do something. Ownership of a patent is not necessary in order to exploit an invention, but neither does it imply freedom to operate the invention. In any developed field of technology, it is likely that there will be underlying technology upon which the new invention builds. If it is necessary to use that underlying technology in order to work the invention, but the underlying technology is still protected by patent or other rights, then the grant of a patent over the new invention does not give freedom to operate. The patent holder will need to obtain a licence of any valid rights over the underlying technology in order to avoid the risk of being sued for infringement. This will be the case for many novel materials and nanoscale structures: the owner of a patent over the material, or the building blocks used by later workers to assemble the structures into novel circuits, products or devices, is able to stop all subsequent commercial use of later inventions unless an assignment or licence can be negotiated. Of course, ownership of a patented improvement on, or application of, the existing technology may well be a valuable bargaining chip in such negotiations. But on its own, it does not entitle its holder to go ahead regardless of other patent rights.

A patent, then, is a purely negative right to stop other people doing that which is within the scope of the patent. This right applies even if they come by that invention independently, without knowing of the earlier work. Potentially, a patent may confer market exclusivity upon its holder. But the commercial effectiveness of any given patent depends both on the scope of the claims – is it a complete road block, or merely an obstacle which with a little effort may be designed around? – and on the validity of the grant. No Patent Office is perfect; thus, a granted patent is not necessarily a valid patent. In any discussion over the value of a particular patent, it is important to come prepared not only with an interpretation as to the true scope of the claims but also as to the strengths and weaknesses of the whole.

A patent is purely a creature of registration; it exists only in those territories where it has been granted and registered, and only for so long as renewal fees are paid. Initial registration is in turn conditional upon the invention meeting the criteria for patentability. At any stage, even after expiry, a patent can be revoked if it is proved that it did not meet one or more of these criteria. If revoked, the patent ceases ever to have had legal effect.

2.2.2 Patentability

2.2.2.1 Legal Principles

Patent law is technology neutral, so that the basic principles which apply to macro-scale engineering, materials and chemistry are equally applicable to nanoscale inventions. It is also an area of law which is considerably more harmonized globally than most other areas of commercial law. This is a result of the success of the WTO, which requires all Member States to subscribe to its Agreement on Trade-Related Aspects of Intellectual Property (commonly referred to as TRIPS). The TRIPS Agreement sets out the minimum criteria which Member States' national patent laws must apply to the grant of a patent:

> Article 27
>
> *Patentable Subject Matter*
>
> 1. Subject to the provisions of paragraphs 2 and 3, patents shall be available for any inventions, whether products or processes, in all fields of technology, provided that they are new, involve an inventive step and are capable of industrial application....
>
> 2. Members may exclude from patentability inventions, the prevention within their territory of the commercial exploitation of which is necessary to protect *ordre public* or morality, including to protect human, animal or plant life or health or to avoid serious prejudice to the environment, provided that such exclusion is not made merely because the exploitation is prohibited by their law.
>
> ...

Thus, it is not open to any Member State of the WTO to refuse to grant patents over nanotechnology unless it can justify so doing on the grounds that the exploitation of the invention would potentially endanger human, animal or plant life or health, or seriously prejudice the environment. So far, the debate around the regulation of nanotechnologies on the grounds of such risks has focused rather on scientific assessment of the risks and benefits rather than any proposal that nanotechnologies should be excluded from patentability in the first place. This is logical, since refusing patents over an invention does not in any way prevent the exploitation of that invention; indeed, it would be more accurate to say that it throws exploitation open to all-comers rather than retaining it (for the duration of the patent, at least) within the control of a single entity whose activities can be supervised.

In practice, therefore, nanotechnology inventions will be patentable, provided that they meet the fundamental criteria of novelty, inventiveness and having some potential industrial application, anywhere in the world. The precise interpretation of each criterion can and does differ from one country or region to another, however, and the discussion below focuses on the European interpretation of these requirements, which are enacted in Article 52(1) of the European Patent Convention (EPC).[3]

An important omission from Article 27 TRIPS is the lack of any definition of what may amount to an invention – and, indeed, there is no positive definition in the EPC either. This omission thus permits Member States to exclude from patentability anything which is arguably not an invention. The relevant exclusion so far as nanotechnology is concerned appears in Article 52(2) EPC which states that (among other things) discoveries and scientific theories shall not be considered to be inventions. At first glance this could be taken to exclude significant new nanostructures or materials which could be said to be discoveries rather than inventions. But this exclusion is very narrow, since it is only the discovery *as such* which cannot be patented.[4] Useful applications of the new structure or material remain patentable, and may be so even if the application is obvious once the discovery has been made.[5] The objective is to ensure that a researcher who discovers a new property but has an interest only in exploiting it in a particular field does not obtain a monopoly across all possible applications which can be used to prevent further exploitation by others until the patent expires.

2.2.2.2 Form of Claims

A patent may claim either a product or a process by which a product is manufactured.

A claim to the product itself can protect the claimed product however made or for whatever purpose it is used, even if the patentee has disclosed only one way to make it and only one specific use. This does not prevent later workers from patenting new uses if the new use reflects a previously unknown property which provides a new technical effect,[6] but such patents will be limited to the specified uses and cannot be exploited without the permission of the owner of the patent to the underlying product. Similarly, a later patent can monopolize a new way of making the product, but it will not be possible to sell the product without the originator's permission. (Conversely, however, the originator cannot decide to use the new process to make the product, or to sell the product for the new use, without a licence from the later patentee.)

Process claims generally result in a narrower monopoly than claims to a product in itself, since a process claim is infringed only by the use of the process. For many products it is possible to devise more than one way to produce them – nanoparticles can be made by traditional wet chemistry, mechanical methods such as attrition ball milling, lithography or vacuum deposition methods and by gas phase synthesis, for example, although all methods will not work for all

3. The EPC is an international treaty separate from the EU, and its membership includes a number of states which are outside the EU including Switzerland and Turkey.
4. Article 52(3).
5. *Genentech's Patent* [1987] RPC 553; [1989] RPC 147, English Court of Appeal.
6. G2/88 *MOBIL OIL/ Friction-reducing additives* Enlarged Board of Appeal of the EPO.

substances – so that the monopoly over the resulting product is effective only for the time that it takes to devise such an alternative. Further, it may be difficult or impossible to prove that a product was made by a particular process, except in those jurisdictions which enable a claimant to obtain detailed information about the alleged infringer's activities.[7] Thus, in general a claim to the product itself is preferable in order to obtain an effective monopoly.

One issue which is likely to arise in nanotech product claims is that of new physical forms of existing elements or compounds. This has already been the subject of considerable discussion in the pharmaceutical context, where an improvement in the stability or ease of manufacturing of a known compound can be extremely useful in establishing a competitive position in the market for that compound. They can also be useful in staving off generic competition when the earlier patent over the compound itself expires. It can, however, be more difficult for a patentee to argue convincingly that a change in form is as significant an invention as the making of the compound or identifying its usefulness, in the first place. For instance, a number of patents over new crystalline forms of known chemicals have been challenged successfully, leading to their revocation.[8]

One final category of claims are known as 'product-by-process' claims – claims to a product defined and produced by a particular process of production. These may be allowed when the product itself cannot be defined in any way other than by reference to its production process, but even when allowed they are liable to be interpreted narrowly when asserted against an alleged infringer. In particular, there is little benefit in bringing such a claim compared with a claim of infringement of a process claim through keeping or selling a product which is the direct product of the process.[9]

2.2.2.3 Novelty and Inventive Step

As stated above, to be patentable an invention must, by Article 52(1) EPC, be new and involve an inventive step. What precisely this means is covered by Articles 54 and 56:

> Article 54 Novelty:
>
> (1) An invention shall be considered to be new if it does not form part of the state of the art.
>
> (2) The state of the art shall be held to comprise everything made available to the public by means of a written or oral description, by use, or in any other way, before the date of filing of the European patent application.

---

7. The European Directive on the Enforcement of Intellectual Property 2001/84/EC and the provision of TRIPs on reversing the burden of proof are discussed at s. 2.2.5 below.
8. Recent examples of such challenges in the United Kingdom include *SmithKlineBeecham v. Apotex Europe* [2004] EWCA Civ 1568; *Les Laboratoires Servier v. Apotex Inc* [2008] EWCA Civ 445.
9. Discussed at s. 2.2.3.2 below.

(3) Additionally, the content of European patent applications as filed, the dates of filing of which are prior to the date referred to in paragraph 2 and which were published on or after that date, shall be considered as comprised in the state of the art.

Article 56 Inventive step:

An invention shall be considered as involving an inventive step if, having regard to the state of the art, it is not obvious to a person skilled in the art. If the state of the art also includes documents within the meaning of Article 54 paragraph 3, these documents shall not be considered in deciding whether there has been an inventive step.

The test for novelty is straightforward: it is simply an objective assessment of whether all elements of the claimed invention have in fact already been disclosed to the public in any single disclosure. Article 54(2), setting out what is considered to be the state of the art, is thus key to the question of patentability.

2.2.2.4 The State of the Art

The state of the art is extremely broad in scope, including any material in any language anywhere in the world, written or oral or in the form of prototypes, models and the like, which is or has been available to the public at the priority date of the application being considered. The priority date is normally the date on which the application was filed at the patent office, although earlier priority can be claimed from a foreign patent application in respect of the same invention if that application has been made within the previous 12 months.[10] It is highly likely that neither the patent examiner charged with assessing whether this criterion has been met, nor the patent applicant itself, will be aware of all such material.

Article 54(3) adds to this vast mass of material the contents of pending but unpublished[11] European patent applications, which the applicant has no means of knowing about – but which will have come to light by the time any dispute over the patent breaks out.

---

10. Article 87 EPC. The earlier application must have been filed in a country which is a party to the 1883 Paris Convention on the Protection of Industrial Property, and have included information such that the whole subject matter of the later application can be derived directly and unambiguously from it, using only common general knowledge in addition to the information provided. If additional features have been added in the second application, then insofar as those features are concerned priority can be claimed only back to the filing date of that second application. G-02/98 *Entitlement to claim priority* [2001] OJ EPO 413. A further complication is presented by the Patent Cooperation Treaty (PCT) system, under which a national application can be used as the basis for a bundle of national patents through a single Paris Treaty filing at the WIPO in Geneva within 12 months of the original application date. A European Patent may be designated as one of such a bundle.
11. Patent applications are not published until 18 months after the filing date, to enable the applicant to withdraw should technical issues come to light during further research which mean alterations to the description and claims may be needed.

The only exclusions from the state of the art for the purpose of European law are for information which became public in the six months before the date of filing due either to being obtained or disclosed in breach of confidence, or being disclosed by display at an accredited international exhibition. Any other disclosure, by the applicant or anyone else, will destroy the novelty of the invention for patentability in Europe and most other territories.[12]

Of course, the more obscure and ephemeral the prior disclosure was the more convincing evidence will be required to prove that it took place. The information in a paper published in a scientific journal will speak for itself, provided that the fact and date of publication can be established. In contrast, the contents of a lecture at a conference must, unless a video recording is available, be inferred from the slides presented or the content of a later-published paper (which of course may not reflect what was actually said on the day, having perhaps been expanded prior to publication). Such inferences will require the support of concrete evidence, such as a statement from someone who attended the conference and is able to say, preferably by reference to contemporaneous notes, exactly what information went along with the slides.

The status as prior art of products allegedly embodying an invention can also be ambiguous. Whilst in theory all information about the product may be available to anyone with the time and analytical techniques at their disposal, in reality only a small number of new products are subjected to comprehensive analysis, and this is likely to be even fewer in an area where analysis requires extremely specialized and expensive equipment, such as is needed to examine structures at nanoscale. Nevertheless, the Enlarged Board of Appeal of the European Patent Office (EPO)[13] considers that provided the information can be uncovered, and reproduced, without undue burden, then it is taken to have been disclosed whether or not anyone actually did take the time and trouble to do so.[14]

A related issue is whether the disclosure of information was 'enabling' in that it was sufficient to enable a worker skilled in the field actually to work the invention. A statement of the existence of a novel nanostructure is unlikely to be enough for this purpose; there must be details of precisely how it is made, unless the literature already describes techniques which the skilled worker would realize could be used for that purpose.[15] Conversely, a disclosure which does not expressly address a problem of nanoengineering, perhaps because it pre-dates the use of the term, can still anticipate an invention if following the directions in the disclosure would inevitably produce a device which would fall within the claimed invention. The standard is a strict one: it is not enough to select specific points from the

12. This contrasts with the position in the United States where a disclosure by the applicant within the 12 months preceding the filing of a patent application is not taken into account for the purpose of assessing novelty. This 12 months is known as the 'grace period', and did at one time lead to a number of American inventions being refused the grant of a patent anywhere other than the United States, but most multinationals and patent attorneys are now aware of the issue.
13. The EPO's tribunal of final appeal.
14. G-01/92 *Availability to the Public* [1993] EPOR 241.
15. T-0206/83 *ICI/pyridine herbicides* [1986] 5 EPOR 232.

teaching of an earlier document and apply them to a particular example in the document to produce something falling within the later claim. This is an adaptation of the earlier teaching, not an inevitable result.[16]

2.2.2.5 Priority

Similarly, a patent cannot claim priority from a document which discloses only the existence of the new material or structure; to be legitimate, the claim must relate back to a document which enables the invention to be practised. Once again, issues may later arise as to the meaning of a document disclosing a novel nanotechnology. If particular knowledge is assumed on the part of the reader which at a later stage proves not to have been commonly known amongst everyone working in the field, then a claim to priority may fail since the earlier document did not enable the reader to carry out the invention without further research. As time goes by this problem will become less acute as a body of knowledge emerges which genuinely is common to most researchers, but for the time being it is essential to take care to spell out all necessary information in detail, either by setting it out within the application document or by incorporating reference to publications where it can be found.

2.2.2.6 The Skilled Person and Common General Knowledge

The question of common general knowledge is also relevant as a key distinction between the test for novelty and that for inventive step. A new technology is inventive if it is 'not obvious to a person skilled in the art' having regard to the state of the art. Obvious simply means 'very plain'. In assessing this, the body of common general knowledge is expressly taken into account, although it is not considered in the more objective comparison undertaken to assess novelty. Thus, a prior description which encompasses all of the salient features of the invention, but leaves elementary technical details unstated, will not anticipate a patent but may well render it obvious since the skilled person would be able to fill in the basic techniques for themselves.

The identity of the skilled person is therefore fundamental, and the information they can draw upon makes the difference between a patentable invention and something which, however dazzling almost all nanotechnology is to the person in the street, is not patentable. The EPO Guidelines[17] define this person:

> The 'person skilled in the art' should be presumed to be a skilled practitioner in the relevant field, who is possessed of average knowledge and ability and is aware of what was common general knowledge in the art at the relevant date.... He should also be presumed to have had access to everything in

16. T-0396/89 *UNION CARBIDE/ high tear strength polymer*[1992] EPOR 312.
17. EPO Guidelines (April 2010) Part C Ch. IV s. 11.3.

> the 'state of the art', in particular the documents cited in the search report, and to have had at his disposal the normal means and capacity for routine work and experimentation. If the problem prompts the person skilled in the art to seek its solution in another technical field, the specialist in that field is the person qualified to solve the problem. ... He may be expected to look for suggestions in neighbouring and general technical fields or even in remote technical fields, if prompted to do so. ... There may be instances where it is more appropriate to think in terms of a group of persons, e.g. a research or production team, rather than a single person.

What the skilled person is not, however, is inventive. Clearly, this is a highly artificial person, since 'routine work and experimentation' in nanotechnology is unlikely to be done by a person without advanced technical skills and equivalent problem solving abilities which to those outside the field would appear highly inventive; but this is why the assessment has to be done against other workers in the same field. The question of what is inventive is not to be answered by considering whether the average person, or even the average person with a technical degree, could do it – by that criterion, more or less all work in very highly specialist areas would automatically qualify as inventive.

The requirement of an inventive step is likely to be a particular issue where silicon chip and analogous technologies such as LED manufacture, which have routinely involved fabricating structures on the nanoscale for many years, are concerned. The introduction of nano-terminology may appear to produce a new range of patentable inventions where in fact what is described has already been done. The same problem has been tackled in macro-scale technologies where an invention may be claimed by introducing a new parameter into the discussion of a known technology. Since none of the prior art specifies this parameter, and human ingenuity can devise credible reasons why it is in some way critical to the operation of the device or process, the patent examiner may be convinced to allow a patent. The national courts, however, are already sceptical of the real level of invention involved in mere changes of scale or other parameters.[18] In many fields it is obvious to want to reduce the size of devices, if for no other reason than the reduction of cost of materials. Accordingly, to be patentable some new property or behaviour resulting from the scale of the structure will need to be shown to prove that there is an invention in reproducing a known device (a mechanical switch, for example) at a nanoscale. Alternatively, if it can be shown that there was some prejudice or specific technical difficulty to be overcome in order to do so, then a patentable invention may have been made.

---

18. 'There is no invention in stipulating a feature which is arbitrary and serves no useful purpose. ... The public are entitled to make obvious products using obvious and ordinary techniques. The selection of a number of these products by reference to an arbitrary parameter which has no technical significance does not involve an inventive step and does not create a patentable invention', *per* Kitchin J in the English Patents Court in *Abbott Laboratories Inc v. Evysio Medical Devices ULC* [2008] EWHC 800.

2.2.2.7 Problems in Examining Nanotechnology Patent Applications

The multidisciplinary skills contributing to nanotechnology research have been a problem for patent examination of early nanotechnology applications, since relevant publications may exist in a wide range of traditional technical fields. Scientific publications far exceed the number of patents, as is typical for emerging technologies; and the rate of increase of number of nanotechnology-related publications accelerated rapidly between 1997 and 2001 and then accelerated further from 2001 to 2006, so that the amount of material to be searched against any new application has expanded enormously in the last decade. Nanotechnology-related information has appeared in subjects from environmental science, organic chemistry and biophysics to the physics of condensed matter and metallurgy. An additional problem resulting from this diversity of technical backgrounds is the lack of standard terminology, so that it may not be apparent to a patent examiner whose own background is in chemistry that a metallurgical paper is in truth describing the same material as that claimed by a patent applicant.

The challenge that this poses for patent examiners, who themselves are likely to have a background in only one technical field, has led to considerable work at the EPO, Japanese and US patent offices to bring together multidisciplinary skills and review publications in their databases across all disciplines, for tagging as potentially relevant. This process of nanotechnology-focused training and organization having begun only in around 2004, it is entirely likely that patents were granted before that date which did not take into account all relevant publications, the examiner not having had the knowledge or documents available. Such patents may, therefore, be more vulnerable to later attack than patents granted after a thorough review of all relevant prior art.

In addition, in 2006 the major patent offices introduced new interdisciplinary classifications for patents relating to nanotechnology in whatever industrial or technical field, so that searching for both prior art and competitor activity in the patent literature is now considerably easier and more accurate. All such patents granted by the EPO will be tagged with the classification Y01N in addition to the more familiar tags for chemistry, materials science and so on; the United States Patent and Trademark Office (USPTO) has added classification 977. This again will contribute to improve the rigour of examination of future applications, and a reduction in the number of potentially invalid patents that are granted, to the benefit of all users of the patent system.

2.2.2.8 Sufficiency

A further important requirement for patentability is laid down in Article 29 TRIPs:

> Article 29
>
> *Conditions on Patent Applicants*
>
> 1. Members shall require that an applicant for a patent shall disclose the invention in a manner sufficiently clear and complete for the invention to be carried

> out by a person skilled in the art and may require the applicant to indicate the best mode for carrying out the invention known to the inventor at the filing date or, where priority is claimed, at the priority date of the application.
>
> . . .

This requirement of sufficient disclosure, reflected in Article 100(b) EPC, is particularly key in any new area of technology and, as has been the case in the last decades in disputes around biotechnology, it is likely to be an important issue for the patents now being applied for and granted over nanotechnologies. In an unexplored field, the skilled person is considered to be prepared to display a reasonable degree of skill in making trials, and to be able to correct obvious errors in the specification if this can readily be done. But 'there must be available adequate instructions in the specification or on the basis of common general knowledge which would lead the skilled person necessarily and directly towards success'.[19] If reasonable skilled addressees can come to different conclusions as to what the patent means, then it is not sufficiently described. Thus, the lack of a universal terminology and accepted body of common general knowledge in the field, together with the fact that knowledge from a number of different technical skills may be needed to carry out the invention, makes this requirement a challenge at the earliest stages of emergence of a discipline. Patent applicants have to make doubly sure that their explanation of the invention will be comprehensible to workers in the field coming from other groups which may have different background understanding and vocabulary.

To sufficiently disclose a new compound or structure, the specification must enable a skilled person to make it. This may require even more than usual teaching given that many nanotechnology inventions are the result of research carried out among a multidisciplinary team from otherwise-unrelated technical backgrounds – microstructural devices, semiconductor physics, organic chemistry. Although the reader of the patent may be a similarly skilled team, there is inevitably more scope for particular key information to slip between the cracks between the applicant's actual knowledge and the readers' assumed knowledge. The application must therefore include full details of all features of the invention, the production or synthesis method, directions as to where any non-standard reagents or materials can be obtained and test methods enabling the skilled worker to verify whether she has achieved the goal.

Where a claim is to a range of related compounds, it is also necessary for all compounds in the range to have the expected advantageous properties, otherwise the claim may be insufficient because it is too broad, in that not everything within the claim solves the technical problem to which the patent is addressed.[20] If performing the invention will require an undue burden of work or actual inventiveness, the patent is insufficient. This is helpfully explained in the appeal decision in *T-1121/03/UNION CARBIDE/Indicator ligands:*[21]

---

19. T-0226/85 *UNILEVER/Stable bleaches* [1988] OJ EPO 336.
20. T-0929/92/AGREVO/Triazoles [1996] EPOR 171.
21. [2006] EPOR 49.

> Where, as in this case, there was a functional definition of a technical feature it was necessary to establish whether or not the patent in suit disclosed a technical concept fit for generalisation which made available to the skilled person the host of variants by such a functional definition .... The two parts of the definition of the indicator ligand given in claim 1 was no more than an invitation to perform a research programme to find suitable indicator ligands which was an undue burden and insufficient disclosure ...

Consequently, a claim to a class of compounds or devices may be insufficient unless the general principle has been identified and explained to the extent that only routine testing is left to be done. This requirement causes continued difficulties since the inventor of a new material which displays particularly desirable properties – for instance, the fullerenes[22] which although made from carbon as a result of their structure at nanoscale display mechanical strength combined with flexibility and electrical conductivity beyond those of known carbon structures – will always be tempted to try to claim it by reference to those properties, despite the fact that there may be materials completely unrelated to the example the inventor has produced which also can meet that description. Such a claim will be invalid, although a claim to the particular material, described in the specification by reference to its composition as well as its properties, will be perfectly sound.

Article 29 leaves to members the option of requiring any patent applicant to disclose the best mode for carrying out an invention. There is no such requirement under European patent law, but it is an important factor in the United States.

2.2.3 The Rights of a Patent Holder

2.2.3.1 TRIPs Requirements

The classes of activity which can infringe a patent or not are stipulated in Articles 28–30 of TRIPs:

> Article 28
>
> *Rights Conferred*
>
> 1. A patent shall confer on its owner the following exclusive rights:
>
> (a) where the subject matter of a patent is a product, to prevent third parties not having the owner's consent from the acts of: making, using, offering for sale, selling, or importing for these purposes that product;
> (b) where the subject matter of a patent is a process, to prevent third parties not having the owner's consent from the act of using the process, and from the acts of: using, offering for sale, selling, or importing for these purposes at least the product obtained directly by that process.

22. CNTs and buckyballs.

> 2. Patent owners shall also have the right to assign, or transfer by succession, the patent and to conclude licensing contracts.

This language makes it clear that these are minimum requirements; signatories can legislate to prohibit other activities in respect of a patented invention.

2.2.3.2 Literal and Purposive Infringement

The basic issue in assessing whether or not a patent has been infringed is how to interpret the scope of the claims. This is a two-stage exercise. The first stage is whether the alleged infringement infringes the claims taken literally to mean exactly what they say. The exercise is straightforward: can every element of the claimed invention be identified in the infringing product? In some cases, ordinarily in chemical and pharmaceutical disputes, such a literal infringement can indeed be identified, particularly since an infringer may be constrained by regulatory requirements to produce a molecule identical to that authorized for sale by the originator. But in almost all other fields it is unlikely that the allegedly infringing product or process will be a literal and exact reproduction of the claim.

The second stage of analysis under European law is, therefore, to assess whether or not the alleged infringement infringes on a purposive construction of the claim. This is specified in Article 69 EPC:

> Extent of protection
>
> (1) The extent of the protection conferred by a European patent or a European patent application shall be determined by the terms of the claims. Nevertheless, the description and drawings shall be used to interpret the claims.
>
> . . . .

Further guidance is given in the Protocol on the Interpretation of Article 69:

> Article 69 should not be interpreted in the sense that the extent of the protection conferred by a European patent is to be understood as that defined by the strict, literal meaning of the wording used in the claims, the description and drawings being employed only for the purpose of resolving an ambiguity found in the claims. Neither should it be interpreted in the sense that the claims serve only as a guideline and that the actual protection conferred may extend to what, from a consideration of the description and drawings by a person skilled in the art, the patentee has contemplated. On the contrary, it is to be interpreted as defining a position between these extremes which combines a fair protection for the patentee with a reasonable degree of certainty for third parties.

Thus, in any dispute concerning infringement of a patent the court has to attempt to ascertain what a skilled person would understand the patentee to have meant by the wording of the claims, taken in the context of the document as a whole.[23] Given that a dispute is unlikely to reach the courts until several, and often many, years after the date when the document was drafted, this is never an easy task.

---

23. Per UK House of Lords in *Kirin-Amgen v. TKT* [2005] RPC 169.

The task is all the more complex where the allegation of infringement concerns a claim to a process where the infringing act is not the carrying out of the process but dealing in a product which is alleged to be the direct product of that process. For instance, in *Monsanto v. Cargill*[24] Monsanto attempted to assert a patent over a process for genetically engineering a plant to resist the herbicide glyphosate (sold as ROUND-UP®) against a cargo of soya bean meal which derived from plants grown in Argentina which had themselves derived from a genetically engineered strain. Although accepting that the ultimate parent plant was the direct product of the process of genetic engineering, and that the meal did contain fragments of the modified DNA used in that process, the English and Dutch courts held that the meal was simply too far removed from the process described in the patent to fall within the claim. The Court of Justice of the EU concurred, ruling that the modified DNA was no longer protected as it was no longer performing its function.[25]

2.2.3.3 Contributory Infringement

In Europe, a category of 'contributory infringement' has been added to the basic forms of infringement, by a number of countries. This form of infringement arises when someone not licensed by the patentee supplies or offers to supply means, relating to an essential element of the invention, for putting the invention into effect knowing or it being obvious to a reasonable person in the circumstances, that the means are suitable and intended for putting the invention into effect. It is complex to bring such a claim since there can be room for argument on several points: whether something is 'means essential' to the invention; whether it was obvious in the circumstances that the means were to be used to put the invention into effect; what was the state of knowledge of the alleged infringer. But it can be a very effective right for breaking the supply chain for complex products. For instance, a supplier of CNTs could be found liable for infringing a patent over a device constructed using such nanotubes, if at the time of supply the supplier is aware of the purpose to which the nanotubes are to be put.

2.2.3.4 Exceptions

Considerable freedom was given to Member States under TRIPs to provide exceptions to the patentee's rights:

> Article 30
>
> *Exceptions to Rights Conferred*
>
> Members may provide limited exceptions to the exclusive rights conferred by a patent, provided that such exceptions do not unreasonably conflict with a normal exploitation of the patent and do not unreasonably prejudice the

---

24. *Monsanto Technology v. Cargill International* [2007] EWHC 2257 (Pat); *Monsanto Technology v. Cefetra BV* (Hague District Court 19 Mar. 2008, unreported).
25. *Monsanto Technology v. Cefetra BV* C-428/08, Decision 6 Jul. 2010.

legitimate interests of the patent owner, taking account of the legitimate interests of third parties.

The most important of these for practical purposes is the exception for experimental use, implemented in most European countries, which permits third parties to carry out experiments relating to the subject matter of a patent without needing permission for the patentee. The exception applies to any experimental work, whether or not the ultimate purpose is commercial, for so long as it remains truly experimental that is, an inquiry into issues around the subject matter of the patent designed to establish information which is not already known. However, uncertainties as to whether this exception protected the clinical trial work carried out by generic pharmaceutical companies to demonstrate the safety and efficacy of their products for regulatory purposes has led to the introduction throughout the EU of a so-called Bolar exemption which excepts from patent infringement any work carried out to generate data for pharmaceutical regulatory purposes and the necessary practical arrangements for that work. Unlike the equivalent exemption in US law, this exemption does not apply to medical devices. Thus, work on a drug delivery system using nanoparticles to deliver a patented drug which would not qualify for exemption under the experimental use exception since the subject matter of the experiments is the efficacy of the delivery system and not the function or properties of the drug itself, would not be exempted under this new exception either. The exception was introduced only in 2004, however, and its scope has not yet been tested in the courts. Uncertainty remains, for instance, as to the position of commercial contract research organizations which carry out work on behalf of a generic company but do not themselves have any intention of applying for regulatory approval. Given that delivery through nanotechnology will almost certainly be carried out by collaborations in which the nanotechnology company will not be the party seeking the regulatory approval to market the finished product, this exemption may rarely be available.

2.2.4 Nanotechnology Patenting to Date

Most patenting to date has been in nanomaterials (generic in nature, enabling a broad range of downstream applications) and nanoelectronics (symptomatic of the traditional 'top down' approach in big electronics companies, where the nanoscale is essentially an incremental development from the pre-existing trajectory of industrial work). The number of patents applied for in instruments – such as the atomic force microscope, the underlying enabling tools for many other fields of nanotechnology – is growing, as well as in the fields of optics, magnetics and chemicals. The more radical 'bottom up' approaches such as self-assembling molecules and machines may still be at the stage of pure academic research not yet reaching the commercial viability necessary before patenting becomes an issue.

The relative youthfulness of the field is reflected in the fact that a substantial fraction of patents are held by universities.

As to the validity of the patents which have been granted, their potential vulnerability to legal challenge has been noted in the discussion of inventive step and sufficiency, above.

2.2.5 Enforcement

2.2.5.1 Proof of Infringement

The most effective form of deterrence is a reputation for determined enforcement of rights. However, compared to enforcing patents over macro-scale inventions, enforcement of nanotechnology patents may be materially more difficult.

The issue is how a patentee can prove that the elements of the claim are present in the allegedly infringing device or substance. In the case of nanotechnology, this is likely to present a challenge for the available analytical techniques. Although it may be possible to reveal the structures present using atomic force microscopy and techniques deriving from it, characterizing the materials in respect of their physical properties may be more difficult. For instance, the operation of devices made up of multiple very thin layers may depend upon characteristics such as the resistivity of adjacent layers. But at nanoscale there may be parallel conducting paths through one or more layers, making the properties of a single layer exceptionally difficult to observe to the standard of reliability necessary for a court to accept.

The courts of common law jurisdictions such as the United Kingdom and United States provide means for a patentee to obtain full disclosure of the manufacture of an allegedly infringing device, which makes such analysis easier. But civil law jurisdictions tend to provide more limited access to a defendant's internal documents, although it may be possible to seize evidence such as samples and computer records for later analysis by a court-appointed expert, to enable at least some conclusions to be drawn. The harmonization of procedures under the European Directive on the Enforcement of Intellectual Property[26] (the Enforcement Directive) aimed at ensuring there should be some means available in every Member State to a right holder to obtain the information and evidence necessary to establish infringement, but the adequacy of those procedures in the special context of high technology was not addressed.

In this context, at least in relation to process patents, the provisions of TRIPs reversing the burden of proof may turn out to be valuable:

> Article 34 Process Patents: Burden of Proof
>
> 1. For the purpose of civil proceedings in respect of the infringement of the rights of the owner referred to in paragraph 1(b) of Article 28, if the subject matter of a patent is a process for obtaining a product, the judicial authorities shall have the authority to order the defendant to prove that the process to

26. Directive 2004/48/EC of the European Parliament and of the Council of 29 Apr. 2004 on the enforcement of IPRs [2004] OJ L 195.

> obtain an identical product is different from the patented process. Therefore, Members shall provide, in at least one of the following circumstances, that any identical product when produced without the consent of the patent owner shall, in the absence of proof to the contrary, be deemed to have been obtained by the patented process:
> (a) if the product obtained by the patented process is new;
> (b) if there is a substantial likelihood that the identical product was made by the process and the owner of the patent has been unable through reasonable efforts to determine the process actually used.
>
> Any Member shall be free to provide that the burden of proof indicated in paragraph 1 shall be on the alleged infringer only if the conditions referred to in subparagraph (a) is fulfilled or only if the condition referred to in subparagraph (b) is fulfilled.
>
> 2. In adduction of proof to the contrary, the legitimate interests of the defendants in protecting their manufacturing and business secret shall be taken into account.

This provision was introduced because a number of countries formerly barred the patenting of pharmaceutical products but did allow the patenting of processes. The effect was that for practical purposes pharmaceutical products were not fully protected, because the key feature of a pharmaceutical product is usually its molecule, and in practice the composition of this is fairly easy to analyse, so that by devising a synthesis method a competitor could relatively easily reach the market. Although the method of manufacture might infringe a granted process patent, the existence of several possible synthesis routes made it extremely difficult to prove infringement. Article 34 attempts to improve the situation by reversing the burden of proof, so that if the alleged infringer has produced an identical product to that produced by the process patent, the onus shifts to the them to show that the product was produced without use of the process covered by the patent. It is no defence in patent law that the alleged infringer independently developed the identical process. Independent creation is a defence only to an action based on trade secrecy.

The reversal of the burden of proof, hence, may be of particular importance in developing countries and economies in transition that did not recognize product patents for pharmaceuticals or in other fields of technology prior to the implementation of TRIPs. With the universal introduction of product patent protection for pharmaceuticals and chemical products, the practical importance of this principle is diminishing in the pharmaceutical field. However, Article 34 may still provide a valuable procedural tool to the developers of new nanotechnology techniques.

2.2.5.2 Remedies

The remedies available for infringement of a patent are: an injunction against continued infringement; damages or alternatively an account of the infringer's profits; an order for delivery up or destruction of infringing goods, and any tools or dies for making them; a declaration that the patent is valid and has been infringed by the defendant; and an order for publication of the judgment, at the defendant's expense.

2.2.5.2.1 Injunctions

The basis for an award of an injunction is that the infringer threatens to continue to infringe unless restrained. The proprietor will normally apply for an injunction as part of the final order at trial, but can also ask for a preliminary or 'interim' injunction at the outset of proceedings, if the effect of the infringer continuing to infringe until trial will seriously damage the proprietor's business.

Interim injunctions are not common, unless the patentee can show that it is likely to suffer damage as a result of the infringement which will not be remediable after trial – for instance, a permanent price depression as a result of the second source of supply being temporarily available. This argument is most often raised in pharmaceutical cases, where the price drop resulting from the launch of a generic product is unlikely to be reversible.

The Enforcement Directive, however, enables an infringer to apply to be permitted to continue to infringe pending trial on the basis of providing a guarantee against any future damages award. There is as yet no case law as to how this will work in practice.

Any final injunction granted after trial will last until the patent expires or is revoked.

2.2.5.2.2 Monetary Relief

Once infringement has been proved, the patentee has to choose what form of monetary compensation they want. One option is damages – to compensate for loss or injury by, so far as possible, putting the injured party in the same position as they would have been in if there had been no infringement. Alternatively, where the patentee might not have been in a position to make the same sales as the infringer has done, and therefore has suffered little or no actual damage, the patentee may instead request an account of the profits the infringer made by virtue of the infringement.

The calculation of damages can be complex, but case law is scant since most actions settle once liability has been established.

If the proprietor has been exploiting the patent by granting licences, then its loss is the capitalized value of the royalties that the infringer would have paid had he taken a licence. This convenient method becomes less easy to use if – as is often the case – the proprietor has granted various licences of the patent at different royalty rates.

The royalty method is not so appropriate if the proprietor does not grant licences but works the patent itself. In this case it is a question of the profits lost as a result of the infringer's activities. Here, many factors may be relevant including the effect on the market place of the infringement, such as:

- whether every sale of an infringing article represents a lost sale for the proprietor (this is assumed, unless the defendant can show otherwise);
- whether the proprietor would have sold other articles along with those lost sales (e.g., the buyer may have also purchased non-patented articles in addition to the patented article, damages for which are often referred to as 'parasitic' damages). These lost sales will also be recoverable provided that the associated damage was foreseeably caused by the infringement;
- whether the infringer had generated additional interest in the patented article through its marketing efforts;
- what the effect on the market was of changing a monopoly into a duopoly (e.g., where the proprietor had to reduce its prices to compete with the infringer);
- typical profit margins for the category of article concerned;
- whether the infringer had deliberately undercut the proprietor's sale price.

On sales which the proprietor would not have made even absent the infringement, the proprietor is entitled to a reasonable royalty. A proper notional royalty rate is that which a potential licensee who had not yet entered the market would pay after a negotiation between willing licensor and willing licensee.

Unlike the US provision for trebling the damages in cases where the infringement is considered to have been wilful, exemplary damages – that is, damages above and beyond the loss suffered by the patentee – cannot be awarded in most European courts.

2.2.5.3 Parallel imports

An issue which arises in Europe is whether or not patent rights can be used to prevent transactions in a patented product after it has been acquired legitimately from the patentee or its licensee. The fundamental principle of free movement of goods is laid down in the Treaty on the Functioning of the EU:

> Article 34
>
> Quantitative restrictions on imports and all measures having equivalent effect shall, without prejudice to the following provisions, be prohibited between Member States.

Article 36 provides an exception stipulating that intellectual property rights (IPRs) are not such a restriction so long as they are not used as a means of arbitrary discrimination or a disguised restriction on trade between Member States. Decisions on these provisions by the European Court of Justice[27] have led to the doctrine of 'exhaustion of rights', akin to the 'first sale' doctrine in the United States: once the rights have been exercised in the first marketing of a batch of goods by the proprietor or with its consent, they are exhausted and cannot be used again against those particular goods.

27. *Centrafarm v. Sterling Drug* ([1974] ECR 1147) and subsequent cases.

Essential to the application of this principle, however, is the idea of voluntary marketing by the patent holder or by others with its voluntary consent. Thus, it will not operate to prohibit a patent holder from relying on that patent to prevent imports where for instance the product has been made and first sold under a compulsory licence.

As a result, a manufacturer of CNTs in one country cannot prevent its customer from selling the nanotubes, once paid for, to another customer in another European country: its patent rights are said to have been 'exhausted' by the first sale.

The opposite rule applies, however, where sales from outside the EU are concerned. A first sale in a non-EU state has no effect on the patent holder's rights within the EU, and accordingly parallel imports of genuine goods from the United States, China or elsewhere can be subjected to customs seizure at the point of entry or be injuncted from further circulation on the market through patent infringement proceedings in the national courts.

2.3 TRADE SECRECY

2.3.1 Secrecy and Nanotechnology

The development of commercial products incorporating nanotechnology has very long lead times, which means that products may arrive on the marketplace only many years after the initial invention, by which time the protection conferred by early-filed patents, with a maximum duration (other than in pharmaceuticals) of 20 years from filing date, may have expired. Further, as discussed above nanotechnology is a good example of a field in which it may not be possible to reverse engineer the structure or composition of the product from the product itself. In these circumstances, funding issues aside it may be worth considering avoiding the significant costs and management time involved in obtaining patent protection and relying instead on keeping the details of the new invention secret. In principle, information can remain confidential indefinitely, unless and until the same information is developed by a third party and released into the public domain, or analytical techniques become available which can reveal the structure and composition of the product from samples on the market. Other than the need to continuously maintain secrecy, there are no costs to relying upon confidentiality to protect research, so it can be a cost-effective mechanism giving longer term protection than patent rights. The principal downside is the risk of a competitor reaching the same invention and patenting it.

2.3.2 What Can Be Protected

Trade secrets are a particular case – material warranting particularly strong protection – of the broader category of confidential commercially valuable information. This is an important category, encompassing many things which fall outside the areas protectable by the statutory IPRs. Classic examples of information which may be protected by confidentiality include technical innovations to a product or process, commercial details such as business plans and customer lists, and the 'know-how' which may be necessary to maximize the efficiency of any kind of industrial or commercial operation.

Whether any particular example of these will qualify as a true trade secret in practice, is a question of fact and degree: pure trivia will not be protected. A good guideline to the principles which will be used to test whether or not information is protected can be obtained from European law governing transactions involving confidential information. The TTBER[28] (discussed further under section 2.5 below) covers licences of know-how, which it defines as 'a package of non-patented practical information, resulting from experience and testing, which is secret, substantial and identified'.[29] For this purpose:

- 'Secret' means that the information 'is not generally known or easily accessible'.
- 'Substantial' means that the information 'is significant and useful for the production of... products'.
- 'Identified' means that the information 'is described in a sufficiently comprehensive manner so as to make it possible to verify that it fulfils the criteria of secrecy and substantiality'.

Similarly, the English Court of Appeal has roughly defined a trade secret as information which is used in the trade or business, whose dissemination is limited or discouraged by the owner, and which if disclosed to a competitor would cause real or significant harm to the owner.[30]

Unfortunately, despite the clear overlap between these definitions, the treatment of confidential information is one of the areas of law least harmonized throughout the EU. Even within the United States it is a matter of state rather than Federal law and therefore varies depending upon the state within which a disclosure takes place. Since it is not possible to address all of the variation here, the discussion below is based solely upon English law and practice.

There is rarely any dispute as to whether ongoing technical research and development qualifies as a trade secret provided that sensible arrangements are in place to limit public access to the information. Rather, the common problem areas are around business information or ideas. However, issues do arise as to the confidentiality of information which may be discussed at scientific conferences unless the conference is one where confidentiality obligations are imposed upon all participants. Once information has been disclosed to more than a few people, and certainly any information which is actually published, for instance by being included in a scientific paper or patent specification, then confidentiality is lost and the information becomes free for all to use – subject to any patent rights.

---

28. Commission Regulation No. 772/2004 Official Journal L 123, 27/04/2004 P. 0011–0017.
29. Article 1(1) TTBER.
30. *Lansing Linde Ltd v. Kerr* [1991] 1 All ER 418.

An important aspect of maintaining confidentiality over key technical information is therefore to ensure that as little as possible is disclosed, and that any necessary disclosures are always made subject to an express obligation to keep the information confidential. Although in theory this can be done by an oral request, evidentially it is far easier to establish that confidentiality had been imposed if a written note or agreement to that effect exists.

2.3.3 Non-disclosure or Confidentiality Agreements

Non-disclosure agreements can be very brief. The only issues which must be covered are: the definition of the information which is to be treated as confidential; the purposes for which the recipient is permitted to use and disclose it onward (if any); and the duration of the obligation of confidentiality.

Of these, the most controversial is the definition of the information. There is a strong temptation to use all-encompassing definitions which claim that any and all information disclosed between two parties is confidential. But in reality the receiving party can more realistically be expected to recognize as confidential and handle accordingly specific identified information, if they are not also expected to treat every other piece of information which may come their way as such as well. Further, use of an all-encompassing definition risks undermining the disclosing party's right to enforce any obligation of confidence at all, by suggesting that the information was not capable of being identified with any specificity. An obligation to keep absolutely all information confidential is not reasonable and is unlikely to be enforceable. The risk can be allayed by incorporating disclaimers to exclude the most obviously non-confidential material – such as information already in the public domain, or already in the recipient's possession, or subsequently acquired from a third party without breach of confidence. Nevertheless, taking care to spell out a clear category of information to be kept secret in the first place can only enhance the likelihood of its secrecy being accepted if a dispute later arises.

An accurate definition is particularly important in the context of pre-licensing or acquisition discussions or other negotiations where there is a real possibility that the parties may not conclude any agreement and revert to the status of competitors. In such circumstances the truly confidential information needs to be ring-fenced in the hands of the recipient to the extent of: restricting disclosure beyond the negotiating team (and any named recipients necessary to carry out any technical evaluation); restricting transfer of the information onto the recipient's computer network; and providing a workable mechanism for the return of all records of the information at the end of the discussions.

2.3.4 Confidentiality and Employees

Information concerning a company's research is necessarily disclosed to the employees and contractors working on the project, and a common source of disputes is the use in a new employment of information obtained from a former employer. It is important to write express confidentiality obligations into any contract of employment or contract for services, since no reasonable recipient could then be left in any doubt as to its intended confidentiality.[31] Even without an express obligation, English law will imply such a term where the company has paid for development work and expenses if it is understood that any inventions made would belong to the company.[32] An express requirement not to *disclose* information may be supplemented by an implied term not to *use* it either;[33] but both should be made express for the avoidance of doubt.

2.3.5 Practical Measures to Maintain Confidentiality

As regards people other than employees, even putting up a sign stating that what can be seen in a given area is private can, sometimes, be enough to impose a requirement of confidentiality.[34] However, in general the more effort an organization has taken to keep its information secret, the better the chance of establishing that it was in fact secret. Information kept on a network connected to the internet, for instance, is necessarily less secure than information on a standalone workstation kept in a locked room. Information kept password protected is more secure than information left in a laboratory notebook on a shelf. Restricting access to those who actually need the information reduces the likelihood of information leaking, whether deliberately or carelessly, to anyone who may release it beyond the organization.

2.3.6 Scope of Rights

2.3.6.1 Breach of Confidence

The rights conferred on the holder of a trade secret under English law are the right to an injunction to prevent unauthorized use or disclosure of the information, and damages for any unauthorized use or disclosure which has already taken place. If, however, the information has already reached the public domain before any injunction is obtained, then confidentiality is lost and it will not be possible to injunct further use and disclosure by third party recipients who were unaware that the information they received was confidential. It may be possible to obtain a limited injunction preventing the original recipient/ discloser from profiting further from their breach for a temporary period, depending upon the precise circumstances.[35]

31. *Dunford & Elliot Ltd v. Johnson & Firth Brown* [1978] FSR 143.
32. *Vestergaard Frandsen A/S and Ots v. Bestnet Europe Ltd and Ots* ([2009] EWHC (Ch).
33. *Thomas Marshall (Exports) Ltd v. Guinle* [1978] 3 WLR 116, [1979] Ch 227.
34. *Shelley Films Ltd v. Rex Features Ltd* [1994] EMLR 134.
35. See discussion in *Vestergaard Frandsen, supra* n. 32.

Further, limited disclosure to a small number of people may not have the effect of destroying confidentiality entirely – either in the complete set of information, or as against the world at large. But the weak point of reliance upon trade secrecy to protect valuable research work is that once confidentiality is lost it cannot be re-established.

2.3.6.2 Derivative Information

Information developed through application of confidential information, but which no longer comprises much of the original information itself, is a particularly difficult category. Since an action for breach of confidence is normally based in equity, it is apparent that the conscience of a person who has used confidential information to derive a new formula or develop a new business plan may still be tainted even though their end product contains none of the original information.

Use of such derivative material will breach the confidentiality of the original material if the derived material can be said to be a disguised embodiment of the original. However, even if the original information is still confidential, it will not always be true to say that the derived material is an embodiment of it; there will be a point of remoteness where the relationship is too tenuous to support such a claim. For example, in *Ocular Sciences Ltd v. Aspect Vision*[36] the English High Court considered the design and manufacture of contact lenses by a company established by a group of former employees of Ocular Sciences. The confidential information concerned comprised their designs and specifications for soft contact lenses, together with procedures for their manufacture and processing including engineering drawings, testing procedures, computer programs, manuals and formulae. Ocular Sciences alleged that the defendants had set up their company to manufacture soft contact lenses to precisely the same designs and specifications thereby using information that had been disclosed in confidence. The defence was that such information as had been used had become part of the employee's skill and knowledge which he was entitled to use for another employer.

The judge accepted that an employee is entitled to deploy his acquired skill and expertise for any employer, though an employer may restrain unauthorized use or disclosure of his confidential information like anyone else. The identification of those items of technology, if any, which the defendants were entitled to use as part of their acquired skill and expertise had raised particular difficulties. However, his conclusion was that whilst details of the former employer's designs were certainly confidential, the new designs had moved beyond merely disguising the original information: they did not incorporate any confidential details from them and were not produced using any process which was exclusive to the former employer in the sense of being more than a variant on known techniques used by a number of competitors in the field. The claim therefore failed. The case illustrates the difficult judgment call which always has to be made in attempting to draw the line between

36. [1997] RPC 269.

a disguised embodiment and information which is simply too remote to found a claim for breach of confidence.

2.4 TECHNICAL STANDARDS

2.4.1 STANDARDS AND THE LAW

Standards are agreed technical specifications which can be used to support safety testing, regulation, procurement (by ensuring a verifiable quality of manufacture, or interoperability between equipment and processes which conform to them) and clear communication. All of these functions are particularly crucial with respect to a new technology such as nanotechnology, and an integral part of developing appropriate legal regulation.

Standards may be formal – developed and approved by a recognized standards body such as the International Standards Organization (ISO) or ASTM International[37] (ASTM) – or informal – developed within a professional institute or trade association for example. They may be international, national or local. Adherence to standards is necessarily voluntary, save where imposed by contract or through adoption by a government agency, but in practice compliance with any standard that is widely accepted can become essential in order to compete effectively on the market. Adoption of appropriate standards, on appropriate terms allowing all market participants to access to any IP necessary to comply with them,[38] can assist in creating a level playing field by reducing the ability of one or a few larger organizations to monopolize any market sector.

The difficulties over definitions and nomenclature, in particular, which present relatively minor issues in other fields have, as discussed in Chapter 1, become a major hurdle to discussion, let alone regulation, of nanotechnology. For example, regulation to prevent environmental damage through the release of nanomaterials, or to ensure their use in foods, medical devices and cosmetics is safe, can only be developed once there are agreed terms to describe which materials are and are not within the scope of the regulation, validated test methods to detect and identify the regulated nanoparticles, and protocols for their toxicity testing. Likewise, occupational safety concerns around exposure to these materials and devices cannot be either evaluated or managed unless there are protocols for the testing methods to be used to assess compliance.

The work of the various committees established to develop standards in these fields is crucial to resolving such difficulties. They do not have the force of law, however, unless explicitly incorporated in legislation. Conversely, compliance with a standard does not necessarily imply compliance with legal requirements.

Since standards must be widely adopted in order to achieve their intended purposes, the process of approving formal standards is one of achieving consensus,

37. formerly known as the American Society for Testing and Materials.
38. Discussed further at s. 2.4.3 below.

although unanimity is not necessarily possible – private commercial interests are often in tension with the objective of agreeing technical specifications which all can use. As a result, the process of adoption of a formal standard often takes several years. Despite this, the political controversy surrounding nanotechnology and the lack of certainty as to what form of regulation might be appropriate meant that a number of international standards had been debated and agreed before any national or regional legislation on the subject was introduced. These are discussed in s. 2.4.2 below.

There is, however, no aspect of public accountability for the standard selected: it is entirely within the discretion of the standard-setting body. Further, standard-setting activities are not normally subject to external approval or review (save to the extent that a standard achieves widespread adoption in practice, or otherwise). The legitimacy of the agreed standard remains dependent upon the range of interest groups participating in the development of the standard. Particularly where international standards are concerned, participation is effectively restricted to those organizations (both public and private) which can afford the resources both to dedicate time and expertise to the activity of standards development, potentially reducing the manpower available for equally pressing administrative or technical work, and to send representatives to international meetings over a period of years. Less developed countries and smaller enterprises are unlikely to be able to contribute to the same extent as intergovernmental organizations and multinational corporations. Accordingly, the adoption of a standard even by a highly regarded international organization such as ISO, is not a panacea exempting regulators at national or regional level from considering the appropriateness of the standard to any given regulatory objective. On the other hand the undoubted technical expertise which goes into such a process and the fact that at least some range of interest groups is involved can result in a standard which provides a useful frame of reference for subsequent legal instruments, resulting in a degree of consistency between different jurisdictions which might not otherwise be achieved.

2.4.2 Standards and Nanotechnology

2.4.2.1 The Need for New Standards

It has been apparent since the late 1990s that new standards were likely to be required as materials and devices moved down to the nanoscale. For instance, the spatial resolution of nanostructures and the chemical sensitivity of assays would both be beyond the standard techniques then in use (even though the metrological standards underlying the SI system are already in use in nanotechnology simply through the invocation of the nanoscale, by reference to the agreed international standard units of length). International standards on chemical analysis, imaging and particle sizing may also be applicable in whole or in part.[39] But since 2003, when China established its United Working Group for Nanomaterials,

39. ISO TCs 201, 202 and 204.

work has been ongoing both at national and international level to agree written standards for the naming, measuring, reporting and testing of characteristics of materials and devices on the nanoscale.

2.4.2.2 Current Standards Initiatives

Following the Chinese initiative, the European Committee for Standardisation (CEN), which is the only European organization recognized in accordance with Directive 98/34/EC for the planning, drafting and adoption of European Standards in all areas of economic activity (with the exception of electrotechnology, as to which CENELEC is recognized, and telecommunications (ETSI)) established a working group on nanotechnology in 2004. CEN/Technical Committee 352 'Nanotechnologies' was established at the end of 2005 to develop a set of standards addressing:

- classification, terminology and nomenclature;
- metrology and instrumentation, including specifications for reference materials;
- test methodologies;
- modelling and simulation;
- science-based health, safety and environmental practices;
- nanotechnology products and processes.

An ISO committee, Technical Committee (TC) 229, was also set up in 2005 and for topics of mutual interest to ISO and CEN, work is being carried out under the Vienna Agreement[40] with ISO lead. To date CEN has adopted five ISO standards, as indicated below.

TC 229's work includes understanding and control of matter and processes at the nanoscale, typically, but not exclusively, below 100 nm in one or more dimensions where the onset of size-dependent phenomena usually enables novel applications; and/or utilizing the properties of nanoscale materials that differ from the properties of individual atoms, molecules and bulk matter to create improved materials, devices and systems that exploit these new properties. Areas in which standardization is in progress include: the mechanisms for testing toxicity of nanoparticles and evaluating their environmental impact; developing measurement techniques and instruments to be applicable at the nanoscale, and calibration procedures and references.

The first concrete result of the ISO's work was the publication of ISO/Technical Standard 27687:2008,[41] which provides terms and definitions related to particles (nanoparticles, nanofibres and nanoplates) in the field of nanotechnologies. It is the first part of a projected series on terminology and definitions documents covering the different aspects of nanotechnologies, and is intended to

40. The agreement on technical cooperation between ISO and CEN, approved by the ISO Council resolution 18/1990 and the CEN General Assembly resolution 3/1990.
41. Nanotechnologies – Terminology and Definitions for Nano-Objects – Nanoparticle, Nanofibre and Nanoplate.

facilitate communications between organizations and individuals both within industry and externally. This was formally adopted by CEN in 2009.

ISO Technical Report 12885:2008, *Health and safety practices in occupational settings relevant to nanotechnologies*, was published in January 2009. It focuses on the manufacture and use of engineered nanomaterials and how to prevent adverse health and safety consequences during the production, handling, use and disposal of manufactured nanomaterials. Since the Report was necessarily based only on then-current information about nanotechnologies, including characterization, health effects, exposure assessments, and control practices, it is expected to be revised and updated as experience of working with nanomaterials increases.

A series of further standards and reports were published in the course of 2010, as shown in Table 2.1. Those adopted by CEN at the time of writing are indicated with an asterisk.

Table 2.1: ISO Nanotechnology Standards 2010

ISO/TS 29701:2010* Nanotechnologies – Endotoxin test on nanomaterial samples for in vitro systems – Limulus amebocyte lysate (LAL) test
38ISO/TS 80004-1:2010 Nanotechnologies – Vocabulary – Part 1: Core terms
39ISO/TS 80004-3:2010 Nanotechnologies – Vocabulary – Part 3: Carbon nano-objects
ISO 10801:2010 Nanotechnologies* – Generation of metal nanoparticles for inhalation toxicity testing using the evaporation/condensation method
ISO 10808:2010* Nanotechnologies – Characterization of nanoparticles in inhalation exposure chambers for inhalation toxicity testing
ISO/TS 10867:2010 Nanotechnologies – Characterization of single-wall carbon nanotubes using near infrared photoluminescence spectroscopy
ISO/TS 11251:2010 Nanotechnologies – Characterization of volatile components in single-wall carbon nanotube samples using evolved gas analysis/gas chromatograph-mass spectrometry
ISO/TR 11360:2010 Nanotechnologies – Methodology for the classification and categorization of nanomaterials
ISO/TR 12802:2010 Nanotechnologies – Model taxonomic framework for use in developing vocabularies – Core concepts

Further vocabulary standards (Parts 4-8) were released in 2011. A range of other projects covering methods for characterization of CNTs have been released.

ASTM International began working on nanotechnology in 2006 and has developed publications on terminology, measuring particle size and various toxicity issues. The International Electrotechnical Commission (IEC) has established IEC/TC 113 to work on nomenclature, measurement and characterization and performance of electrical and electronic products and systems at nanoscale.

National committees are also working on the issues in most developed nations.

It goes without saying that the list of standards potentially applicable to any new nanotechnology-enabled product will continue to expand rapidly in the next few years.[42]

2.4.3 Standards, Patent Rights and Competition

Collaboration in standards organizations which includes representatives from several commercial interests can be uncomfortably similar to prohibited, anti-competitive collusion.[43] Collaboration efforts generally can be carried out in a manner consistent with competition laws. Organizations should make sure, however, that they are complying in substance and not just in form. This mistake was made by Phillips in a patent pooling agreement which it intended would smooth the market penetration of its video cassette recorder (VCR) format against competition from Sony's alternative. The agreement recited a pro-competitive objective: to increase consumer choice by making available to consumers video cassettes using the Phillips standards from a range of different manufacturers. However, consumers were not to have the benefit of a choice of format from those same manufacturers: the agreement not only prohibited the parties from making any video cassettes incompatible with the format but also from distributing such cassettes made by other manufacturers. Competition between products was reduced. In addition, a party which decided to leave the pool immediately lost the benefit of the licences it had under the remaining parties' patents, but could not withdraw the licences those remaining parties had under its own. Effectively, any company which signed up would not withdraw, since to do so would mean throwing away all the investment it had made in developing compatible technology in order to start afresh with a different format. Competition between technologies was therefore suppressed. The agreement was in effect for less than two years before indications as to the European Commission's views caused the parties to withdraw it – which did not, however, prevent the Commission from completing its investigation and ruling that the agreement had breached European competition principles.[44]

The distinction illustrated in the Phillips case between product markets and technology markets is an important one: both need to be considered before a company can be confident that its agreements are not anti-competitive. Indeed, in the Guidelines to the current TTBER[45] (the Guidelines) the Commission also

42. A technical discussion of the standards in place by 2010 is given in Murashov and Howard, *Nanotechnology Standards* (Springer, 2011).
43. Competition law is discussed more broadly in connection with licensing in general, in s. 2.5 below.
44. [1978] 2 CMLR 160.
45. OJ C 101/2 27.04.04.

refers to the third arena: the market in innovation, the impact on which must also be taken into account.

The essential distinction is between collaboration which attempts to deal with market failures so as to make the market work better, and joint efforts which attempt consciously or unconsciously to supplant market competition. When examining such arrangements, the enforcement agencies' essential objective is to determine whether the collaborative efforts will create efficiencies and promote competition or stifle competition and work to the detriment of consumers. In stark contrast to the VCR format case, the Commission decision on the Moving Picture Experts Group's MPEG-2 Licensing Programme shows that competition principles can operate in favour of collaborative arrangements provided that these are properly formulated. The Commission concluded that an arrangement whereby certain holders of patents essential to the implementation of the MPEG-2 standard agreed to license their patents through a single, non-exclusive and non-discriminatory licensing programme allowed the quick and efficient dissemination of the MPEG-2 technology – to the benefit of consumers.[46] The appointment of a single US undertaking to act as licensor and grant parties interested in exploiting the new technology a Patent Portfolio License giving access to all of the parties' patents through a single licence did not act to give that administrator undue market power because of the terms upon which the arrangement was made.

It is nonetheless clear that the fact that patent pools and jointly agreed standards *can* be pro-competitive does not mean that they will always be accepted to be so. The TTBER expressly excludes agreements involving more than two parties, which means that all multi-party arrangements have to be considered individually and their pro- and anti-competitive effects weighed in the balance. Despite their exclusion from the TTBER, technology pooling arrangements are discussed at some length in the Guidelines. Technology pools, as opposed to formal standards fixing, may in practice amount to a price fixing cartel if the pool consists solely or predominantly of substitute technologies. The establishment of a pool can also support an industry standard or establish a new standard, thereby reducing competition in the market for innovation by foreclosing alternative technologies from the market. The upsides, creation of a 'one-stop shop' for licensing, reduction of transaction costs and setting limits on cumulative royalties (a real issue in other heavily IP protected industries such as biotechnology) are pro-competitive.

The balance between these two sets of factors depends on whether the technologies in the pool are substitutes for each other. If so, the Commission's view is that the royalties are likely to be higher than they would be in the absence of a pooling arrangement since a prospective licensee would otherwise only pay for one of the options. In fact it may be difficult to draft arrangements for licensing a pool consisting substantially of substitute technologies which is not viewed as anti-competitive. Alternatives such as leaving the contributing parties free to license

46. Decision IV/C-3/36.849.

their own portfolios independently will not suffice; the Commission does not believe they will have a sufficient incentive to do so in reality.

On the other hand, if the technologies pooled are complementary, each being required in order to manufacture a given product, the Commission's view is that the arrangement reduces transaction costs and may lead to lower royalties overall. They will therefore generally be pro-competitive in effect, especially if each of the patents concerned is essential to the manufacture of a given product – even if the participants wield significant market power.

Notably absent from the guidance given in the Guidelines is any consideration of parties' conduct in settling upon the patents to be included in the pool, except for the expression of a preference for independent experts to be involved or at the least for the selection 'panel' to include representatives of all interest groups. This may change when the next revision of the Regulation and Guidelines is issued as the Commission is becoming more concerned. The issue was brought to the fore by the *Rambus/ JEDEC* disputes, where the US Federal Trade Commission (FTC) charged Rambus, a US memory chip technology company, with violating antitrust provisions by withholding information about its patent applications from the Joint Electron Device Engineering Council (JEDEC) standard-setting group for solid state electronics until after a standard had been agreed and then claiming royalties from chip producers. The specific issue complained of was Rambus' alleged failure to disclose certain work in progress to the standards committee working on the Synchronous Dynamic Random Access Memory (SDRAM) standard. In fact Rambus participated in JEDEC only for a short period. It first attended a meeting December 1991 but withdrew in June 1996, citing in particular that its proposed technology licensing terms might not be consistent with JEDEC's terms.

At the outset of this period, Rambus had pending before the USPTO a number of applications, all divided out from or partial continuations of an original application it had filed in 1990. None were granted in the form they had held in 1991, and several underwent further amendment before their ultimate grant. However, once certain of the patents had been granted Rambus asserted that they covered technology required by the SDRAM standard and demanded manufacturers pay royalties for their use. Since Rambus had not submitted these patents to the licensing terms approved by JEDEC, its action was characterized as a 'patent ambush'.

The issues were not, however, entirely clear cut. In particular, JEDEC's terms of participation were not precise as to what disclosures were or were not required of members. Even in giving evidence to the FTC, former participants could not say for certain whether a company which had a patent application pending over technology on which it was continuing to work had any obligation to disclose that ongoing work in case the ultimate scope of a granted patent was pertinent to the standard. Nor was it clear what level of belief a participant needed to have as to the likelihood of a technology being included in the standard before the obligation to disclose would arise.

The FTC brought proceedings in 2002 for violation of the Sherman Act, the fundamental US law on antitrust. It claimed that Rambus' failure to disclose its potentially broad pending patent applications in the course of the standardization work was a breach of JEDEC's policy and thus amounted to unfair competition and unfair or deceptive practices. Although the charge was initially dismissed, the FTC on appeal concluded that Rambus had indeed deprived JEDEC of one of two options:

(i) it could either have chosen to include different technology in the standard had it been aware of Rambus' intellectual property; or
(ii) it could have required Rambus to accept licensing terms on a reasonable and non-discriminatory basis *before* the standard was finalized.

Although the FTC could not reach a conclusion on the available evidence as to which of these JEDEC might have done, it took the view that either way competition had been harmed. Rambus' deception of JEDEC had significantly enhanced Rambus' acquisition of monopoly power.

On appeal, the Court of Appeals for the Federal Circuit found the evidence insufficiently conclusive and overturned the FTC's ruling.[47] But equivalent proceedings brought under European competition law[48] in 2005 were settled in 2009 when the European Commission adopted a decision making the commitments offered by Rambus, to cap royalty rates for these products, legally binding. This included Rambus agreeing not to charge royalties for its patents relating to the SDRAM and DDR DRAM standards, which were adopted during the time in which the Commission considers Rambus may have been carrying out a patent ambush.

In the wake of this, the Commission has required changes to the European Telecommunications Standards Institution's (ETSI) IPR policies to minimize the potential for companies to exploit the body's rules to pull off similar patent ambushes, and ensure that standards are agreed according to transparent criteria. Further, in December 2010 Guidelines were adopted setting out various criteria (the 'safe harbour') against which a standard-setting agreement may be considered acceptable.[49]

The safe harbour criteria require that the procedure for adopting the standard is unrestricted with participation open to all relevant market competitors and is transparent, so that stakeholders can inform themselves of upcoming or ongoing work. There must also be a balanced IPR policy with good faith disclosure of all IP rights which may be essential to the development of the standard and an irrevocable commitment from IP right holders to offer to license their essential IPR to all third parties on fair, reasonable and non-discriminatory (FRAND) terms. If a patentee fails to offer such terms in respect of patents declared essential to such a standard, this can be used by an alleged infringer as a defence to an allegation of patent infringement.

---

47. *Rambus Inc. v. FTC*, 522 F.3d 456, 466–67 (D.C. Cir. 2008), *cert. denied*, 129 S. Ct. 1318 (2009).
48. Article 102 Treaty on the Functioning of the EU (Lisbon Treaty).
49. Guidelines on the applicability of Art. 101 of the Treaty on the Functioning of the EU to horizontal co-operation agreements OJ C11, 14 Jan. 2011, 1 (the Horizontal Agreement Guidelines).

The lack of independent review of industry sector standards can mean the system is effectively trust-based, so that any participant may declare numerous patents essential without discussion as to whether or not the standard can be implemented by technology not incorporating the inventions protected by those patents. This can be problematic since manufacturers then risk being sued for infringement if they claim that their products comply with the standards without taking a licence, despite holding the view that the patents are not genuinely necessary to meet the standard. The English courts have dealt with this by introducing a new remedy, a 'declaration of non-essentiality',[50] to enable an organization that is being asked to pay licence fees in respect of patents which it does not believe truly are necessary in order to produce a product compliant with a given standard to obtain a court ruling effectively carving that patent out from the standard.

2.5 LICENSING

2.5.1 Practical Issues in Bringing a Nanotechnology Product to Market

The challenges faced in commercialization are commonly identified as a material issue by surveys of companies working in nanotechnology.[51] High processing costs, problems in the scalability of research and development for prototype and industrial production, the basic research orientation of the related sciences, as well as the nebulous concerns around environment, health and safety issues all affect companies' ability to move their ideas into commercial exploitation and begin to generate revenues to support further research activities. Further, the basic research underlying new nanotechnologies has a lengthy lead time before the true prospects for commercialization can realistically be assessed let alone achieved, requiring sources of funding other than those of venture capital or private equity with their usual focus on short term return on investment.

This is particularly frustrating given the proliferation of studies forecasting very rapid growth of the markets for nanotechnology-enabled products. Estimates of USD 12–13 billion in 2007–2008 are expected to increase to hundreds and even thousands of billions by 2015.[52] Of course, such figures are critically dependent on which technologies and products are included within such markets by their authors – is it the industrial market purely for supplies of nanomaterials, or the consumer market for all products incorporating such materials for instance? The most optimistic figures are therefore almost certainly over-inflated. Nevertheless,

50. *Nokia Corporation v. Interdigital Technology Corporation* [2005] EWCA Civ 614.
51. Palmberg, Dernis & Miguet, *Nanotechnology: an overview based on indicators and statistics* STI Working Paper 2009/7, OECD Directorate for Science, Technology and Industry.
52. See forecasts published by LuxResearch, BCC and Cientifica in 2008.

the emergence in the market of actual nanotechnology-enabled products within the last few years should add credibility to the claimed potential.

In addition, given that a major portion of nanotechnology research is still taking place in university departments and spin-offs or small start-up companies, managerial and organizational capacity issues can cause difficulties in moving towards commercialization. The need to create networks with other laboratories, shareholders and other funders, to handle regulatory concerns and to access capital, manpower and the appropriate skill-set as well as the production infrastructures needed can all limit an enterprise's ability successfully to develop a viable product.

One approach which may alleviate some of these difficulties is the sharing of technology through licensing. Granting licences over a technology at an early stage of development can contribute to the income necessary to further development, while more complex agreements may include access to another company's technology, expertise and facilities in return for sharing the benefits. Other than purely publicly funded enterprises or research taking place within an established company which already has products on the market providing an income stream, a large number of nanotechnology companies will at some stage in their development need to enter into licensing arrangements.

2.5.2 Formalities

Patents and patent applications may be assigned or licensed, either before or after grant and either exclusively or otherwise. They may also be mortgaged or vest by operation of law in a personal representative or trustee in bankruptcy, for example, like other personal property. The formalities applicable to each of these transactions are purely a matter of national law.

Under English law, for instance, an assignment or mortgage of a patent or patent application is void unless it is in writing and signed by or on behalf of the patent owner.[53] All transactions affecting property rights in patents, including exclusive licences, must be entered on the Register of Patents. Failure to register means a potential loss of priority over any later assignee or mortgagee without notice of the earlier transaction who registers first.[54] There may also be financial penalties should it later be necessary to enforce the rights.[55] These follow-up actions are a frequent cause of errors in the chain of title of patent rights, as the parties lose interest once the transaction has been completed. It is essential to include in any transaction disposing of patents a covenant of further assurance to ensure that if, on a later assignment, it transpires that the previous assignment has not been registered, the former owner can be required to assist in setting the record straight. However, obtaining the practical co-operation necessary for this several years after the transaction took place tends to be difficult.

53. Section 30(6) Patents Act 1977.
54. Section 33 Patents Act 1977.
55. Section 68 Patents Act 1977 (as amended).

Trade secrets, on the other hand, are not a property right at all under English law.[56] It therefore is not possible to 'assign' or mortgage rights in trade secrets; but the information can still be disclosed in the context of an agreement which determines what use the recipient can make, for how long and for what purpose. The discloser can agree not to use the technology for the same purposes and the same period. Such an agreement therefore has many of the same characteristics as a licence under patent rights. There is, however, no requirement for the agreement to be in writing[57] nor any public register upon which the agreement needs to be entered.

2.5.3 Key Terms

The terms appropriate for any technology licence are specific to the particular technology and the proposed relationship between the parties; there is nothing exceptional about a licence which relates to nanotechnology. The topic is therefore largely beyond the scope of this book, but the following is a checklist of some key terms to consider.

2.5.3.1 Grant

This should define the scope of rights being granted under the patents and applications being licensed, focusing on what rights the licensee wants or needs in order to use the technology, rather than merely encompassing all available rights. The rights granted may be limited to a particular field of activity, or territory; to manufacture or sale. They may be 'exclusive', meaning that the licensor itself will no longer be able to use the rights in that field or territory, 'sole' such that the right holder and licensee share the right to work the invention but no further licences can be granted, or 'non-exclusive' such that the licensor can continue to use and to grant further licences to third parties.[58]

The parties should also consider whether related confidential know-how, not included within published documents, should be licensed along with the patent rights.

2.5.3.2 Maintenance and Policing

Since patents continue in effect only so long as the renewal fees are paid, a licence should specify which of the parties is taking on responsibility for paying them.

56. The contrasts strikingly with the law of some US states which have expressly conferred property status on information – so-called proprietary information.
57. Although it is certainly best practice to do so.
58. A further contrast with US legal terminology in which the phrase 'sole and exclusive' is used to convey the meaning carried by 'exclusive' under English law.

Similarly, the value of a patent which is seen by competitors to be infringed with impunity will rapidly become disregarded by all. One or both parties should undertake responsibility for policing the market – at the least, the licensee should be obliged to report any potential infringements in its marketplace of which it becomes aware.

2.5.3.3 Technology Transfer Mechanism

Unless the licence is a bare licence conveying the right to work the invention and nothing more, the parties need to agree how any additional material is to be transferred – in documentary form, through provision of samples or tools or even through training or consultancy arrangements. If such additional know-how is being transferred, the licence should also stipulate that it is to be kept confidential.

2.5.3.4 Improvements and Rights in Joint Development

If the licensee will be carrying out further research and development such that it may devise improvements to the invention, the licence should address what rights (if any) the licensor is to have over those improvements. This can, depending on the market position of the parties, include obligations to assign or exclusively license the licensee's severable improvements to the licensor. An approach which is less likely to raise competition law issues[59] is an obligation on the licensee to grant a non-exclusive licence to the licensor; such an obligation may, but does not have to be, reciprocal so that any improvements the licensor makes during the term of the licence are similarly licensed to the licensee.

Where the parties intend to work together to develop products, new patentable inventions may arise. It is essential to consider how the rights in such development will be handled before the work is actually done, since the potential for disputes is heightened once a valuable asset is in play. Whilst joint ownership is superficially attractive, it can cause problems in practice since national laws deal with jointly owned patents very differently. For instance, under US law each joint owner has the right to exploit a patent independently, including by assignment of its share to a third party, but English law requires all joint owners to consent to any transaction in the patent including licensing. Once the parties have fallen out, such consent may be impossible to obtain. Accordingly, it is prudent to spell out sole ownership by one party but appropriate licences in favour of the other party, enabling it to exploit the technology as required for the future.

2.5.3.5 Payment

The arrangements for payment also depend entirely on the stage of development of the technology and the relationship between the parties. A one-off licence fee may

59. See s. 2.5.4 below.

be paid up front, there may be milestone payments as specified stages of further development are reached or there may be running royalties. Where royalties are proposed, it is critical to define the basis for their payment, bearing in mind that patent applications pending when the licence is signed may not grant, or may grant only in a limited form. Similarly patents may expire and know-how may become public. The licensee may well not wish to continue paying royalties in such circumstances, but unless the licence clearly allows for such cessation, a contractual obligation to pay may continue.

In its decision in *Oxonica Energy Limited v. Neuftec Limited*[60] the English Court of Appeal considered a pair of related agreements concerning the joint development of a fuel additive using nanoparticles which nicely illustrate these issues. Before beginning development the parties agreed a patent and know-how licence on basis of a pending Patent Cooperation Treaty (PCT) patent application which, as is usually the practice with international patent applications, was in very broad terms.[61] It was also capable of maturing into a granted patent in all of the countries parties to the PCT – but not elsewhere. The licence was to give Oxonica worldwide rights to make and sell Licensed Products, defined as those falling within the claims of 'the Licensed Application or Licensed Patent'. Oxonica accordingly developed and launched a fuel additive, Envirox, using the combined technology.

Predictably, the patents later granted within the EU were narrower than the PCT application had been, specifying a particular coating material to be used for the nanoparticles. Oxonica's initial product, Envirox, fell within both versions. They later developed Envirox II which fell within the PCT claims but not the granted European Patents. Oxonica therefore refused to pay royalties on Envirox II, arguing that 'no one could interpret the licence to say they should pay royalties when there was no granted patent covering the product'. They would be facing legitimate, unlicensed competition and be at a price disadvantage. They also argued that the fact of the worldwide know-how licence could, in practice, be disregarded since once the PCT application was published there could be no further licensable know-how – all had been made public.

The Court of Appeal ruled that royalties were payable on all sales, even in countries where the product was not within the scope of the granted patent claims. In this case, the context required that both the application and any ensuing patents should be licensed, but given the wording it was not necessary to tie the payments to exactly what was being licensed as the payments also related to know-how, sales in countries where there was no application pending, and in addition a non-competition clause. As to Oxonica's point that no know-how remained after publication of the PCT application, on the face of the agreements themselves it had to be taken that the parties anticipated a material body of know-how being provided in addition to the patent information and rights.

60. [2009] EWCA Civ 668.

61. Normal practice in patent applications is to claim the invention as widely as possible, and then progressively retreat to narrower claims as the Patent Examiner raises specific objections.

Payment clauses should also specify at what point royalties become payable, which may be immediately, when a commercial product launches or at another time.

Finally, the calculation of amount payable must be precise, bearing in mind also where international agreements are concerned that withholding tax may be payable.

2.5.3.6 Duration and Early Termination

The parties should consider whether the licence can be terminated on a no-fault basis and if so the appropriate break points. It is also conventional to allow for termination in respect of specified events such as the insolvency or takeover of the licensee. Termination for breach of a material term can also be included, or for failure to reach particular milestones (which must be well defined), or if the technology becomes obsolete in view of developments in the market.

The licence should also set out the consequences of termination, such as what ongoing obligations of confidentiality apply and what is to happen as regards return of materials provided by the licensor. If the licensee has permission to grant sub-licences, the licence should specify what happens to these if the head licence terminates.

2.5.4 Competition Issues

European competition law makes void any contracts or other agreements which may affect trade between Member States and which have as their object or effect the prevention, restriction or distortion of competition within the common market.[62] The individual Member States have equivalent provisions as to purely domestic trade, in their national laws. There is no need for the parties to intend to restrict competition; the impact of the agreement is considered objectively. Since patents provide an absolute monopoly to their holders, there is frequently a risk that any transaction with either suppliers, customers or potential licensees could infringe one of the prohibitions on restricting or distorting competition laid down by EU and national law. If so, the agreement is unenforceable, unless exempt, and the parties may be liable to a fine of up to 10% of worldwide turnover.

Competition analysis applies wherever the parties to an agreement have a combined market share of more than 10% of the relevant market. But this merely begs the question of what should be included in the relevant market, a question of economics. Even for agreements between holders of small market shares, there is the risk that the agreement will be so successful that the parties' market share increases and eventually exceeds 10%. It is therefore prudent for any but the smallest companies to consider the potential impact of competition law on any patent or know-how agreement.

---

62. Article 101(1) Treaty on the Functioning of the EU.

Clauses likely to be found in breach of Article 101 include: terms limiting a party's geographical scope of operations or sales; 'tying clauses' requiring one party to obtain unpatented articles or materials from the other; and restrictions on the right to terminate. Other terms that might be caught are: 'no-challenge clauses', which prohibit a party from challenging the validity of a patent; terms limiting the numbers of articles to be made; prohibitions on exports by either party; and terms requiring payment of royalties beyond the expiry of the rights concerned, although this does not restrict ongoing royalties if know-how is also being licensed. As discussed at section 2.4.3 above, patent 'pooling' arrangements may also conflict with Article 101.

Agreements prohibited by Article 101(1) may still be valid and enforceable under Article 101(3), which states

> The provisions of paragraph (1) may, however, be declared inapplicable in the case of [any agreement or concerted practice] which contributes to improving the production or distribution of goods or to promoting technical or economic progress, while allowing consumers a fair share of the resulting benefit and which does not:
> (a) impose on the undertakings concerned restrictions which are not indispensable to the attainment of these objectives;
> (b) afford such undertakings the possibility of eliminating competition in respect of a substantial part of the products in question.

Thus, a balance needs to be assessed between the extent of possible market distortion by the agreement and the resulting benefit to consumers.

The number of patent licences notified to the European Commission seeking exemption from Article 101[63] rapidly became too great for it to be able to examine each one even cursorily. It was also having to consider many of the same terms in each licence notified. The Commission has therefore issued a series of Block Exemptions. There are currently four Block Exemptions relevant to IP, each applicable in a particular context and accompanied by detailed Guidelines as to how it is applied. The TTBER applies to most patent and know-how licences other than those which are granted for the purpose of research or development of new products. Agreements having this purpose are exempted (if applicable) under the Research and Development Block Exemption.[64] Further categories of agreements having some IP provisions will be exempted under the Vertical Agreements Block Exemption,[65] which relates to agreements for the supply of goods and services between companies in a single supply chain from manufacture to sale to the end

63. Strictly speaking, its earlier incarnations.
64. Commission Regulation (EC) No 1217/2010 on the application of Art. 81(3) of the Treaty to categories of research and development agreements (OJ L 335/36, 18 Dec. 2010).
65. Regulation 330/2010 of 20 Apr. 2010 on the application of Art. 101(3) of the Treaty on the Functioning of the EU to categories of vertical agreements and concerted practices. OJ L 102, 23 Apr. 2010, 1–7.

user, and the Specialisation Agreements Block Exemption,[66] relating to agreements where one party agrees to cease production of one type of product in favour of buying it from the other and joint production agreements.

Whether an agreement will be exempt under the applicable Block Exemption depends upon an analysis of the potential effects of the agreement on the relevant market, a task involving careful economic analysis beyond the scope of this book. Where an agreement does not fall within any Block Exemption, then the Horizontal Agreement Guidelines should be taken into consideration in negotiating any potentially anti-competitive agreement. Competition law is a highly specialist subject, and in practice it is likely to be necessary to take specialist advice.

2.6 ADVERTISING

The absence of any clear information about the actual use of nanomaterials in consumer products, and the use of the term 'nano' to advertise products which may not contain any nanomaterials combine to make it impossible for consumers who are concerned about nanotechnology to select products accordingly. There are no regulations specifically relating to the advertising of products including nanomaterials, but the general principles of advertising law apply.

Directive 2006/114/EC concerning misleading and comparative advertising[67] (the Misleading Advertising Directive) defines as 'misleading' any advertisement which:

> in any way, including its presentation, deceives or is likely to deceive the persons to whom it is addressed or whom it reaches and which, by reason of its deceptive nature, is likely to affect their economic behaviour or which, for those reasons, injures or is likely to injure a competitor.[68]

In determining whether an advertisement is misleading, by Article 3 account is to be taken of the composition of the goods which could include any claims concerning the presence or absence of nanomaterials. However, enforcement is a matter of national law and although Member States are not prohibited from giving consumers a right to enforce against advertisers directly, the Directive itself contemplates instead that people having a legitimate interest in combating misleading advertising or regulating comparative advertising may take legal action against such advertising; or bring such advertising before an administrative authority competent either to decide on complaints or to initiate appropriate legal

66. Commission Regulation (EU) no 1218/2010 of 14 December 2010 on the application of Art. 101(3) of the Treaty on the Functioning of the European Union to certain categories of specialisation agreements (OJ 335/43 OJ 18.12.10).
67. OJ L 376, 27 Dec. 2006, 21.
68. Misleading Advertising Directive Art. 2(b).

proceedings. Whether consumer groups are considered to be persons have a legitimate interests is up to each Member State to decide, when implementing this provision.

Of course, whoever is entitled to bring proceedings is unlikely to be able to do so unless able to analyse the product to establish whether or not any 'nano-claims' are true. Further, the Misleading Advertising Directive is unlikely to apply where no information is given as to the presence or absence of nanomaterials, since if no such information is required to be given it can hardly be argued that its absence is misleading. Consumer groups and the European Parliament, in its 2009 Resolution, have therefore called for the introduction of a requirement of product labelling if nanomaterials are used.

At the time of writing, however, there has been no proposal for any amendment to the Misleading Advertising Directive. Labelling requirements being introduced in respect of cosmetics and food are discussed in Chapter 6.

Further legislative provisions controlling advertising content at the point of sale (i.e., on products or material supplied with products) are governed by the Unfair Commercial Practices Directive 2005.[69] This Directive, and its national implementations, protect consumers against unfair trading practices. A practice can be unfair generally, or can be unfair by virtue of being a misleading action or omission, or by being unduly aggressive.

Commercial practices are misleading if they give false information which deceives (or is likely to deceive) the average consumer and causes or is likely to cause him or her to take a transactional decision that he or she would not otherwise have taken in relation to: (i) the nature of the product; or (ii) the main characteristics of the product. The main characteristics of the product include the specification of the product.

Thus, for instance, a statement that a product or its packaging incorporates nanotechnology is a statement in relation to the specification of a product. The statement must be likely to cause the customer to make a purchase. This must surely be the intention (the implication being that nanotechnology is an added benefit of the product or packaging).

If the statement is false, in that the product and its packaging do not contain any nanotechnology, then such advertising could well be found to constitute an unfair misleading practise. It is not surprising, however, given the relatively small number of products on the market in the EU either containing nanomaterials or being advertised as doing so, that there have been no such cases to date.

69. Directive 2005/29/EC OJ L 149/22 November 2005, 6.05.

Chapter 3

Nanotechnology in Unregulated Sectors

3.1 OVERVIEW

Although the legislators' and regulators' ultimate intent may be to ensure that no nanotechnology products reach the market which are unsafe to use or capable of causing harm through their manufacture or disposal, the complexities of the global market and supply chain are such that no legislative or regulatory approach will be entirely comprehensive, and certainly not for some time to come. This chapter therefore outlines the legal mechanisms other than breach of specific regulatory regimes which could give rise to liability for a manufacturer of consumer products incorporating nanomaterials or the suppliers of such materials, in Europe.

3.2 PRODUCT LIABILITY LEGISLATION

3.2.1 The Product Liability Directive

The EU has taken a significant role in legislating for the safety of consumer products. The first such legislation was the adoption in 1985 of Directive 85/374/EEC on liability for defective products (the Product Liability Directive).[1] This provides that the producer of a product will be liable for damages for any personal injury and certain forms of damage to property caused by a defect in the product, without the person injured needing to show any fault (such as negligence) on the part of the producer. If the producer cannot be identified, and the person who supplied the product fails to inform the injured party who their supplier is within a

1. OJ L 210, 7 Aug. 1985, 29–33.

reasonable time, whether or not the injured party specifically asks for the information, then the supplier becomes liable in the producer's stead.[2]

'Producer' for this purpose includes someone who adds their own brand to the product without having manufactured it, or the person who imports it into the EU.[3] The objective is to make sure that a consumer who is injured has someone from whom to seek redress without having to go outside the Union or carry out intensive detective work to identify the producer.

Products covered by the Directive include all movables, which means that components and substances for incorporation into other products are included. The Directive therefore applies both to nanomaterials and to nanodevices, such as molecular scale switches, once these reach commercialization.

The category of property damage which can be compensated under the Directive includes any product or property worth more than EUR 500, excluding the defective product itself, provided that the property is of a type ordinarily intended for private use or consumption and was used by the injured person mainly for his own private use.[4] Thus, if the defect is in a filter which is supplied as part of a domestic washing machine and causes damage to the washing machine and other adjacent electrical equipment, the consumer can claim damages for the adjacent equipment but not for the washing machine itself. If, however, the filter was a replacement for one which had originally been in the washing machine, then it would be arguable that the washing machine itself could also be recoverable. However, if the washing machine were an industrial one for use in hospitals or care homes, then none of the damage would be recoverable since it would not pass the test either of being intended for private use or in fact mainly privately used.

A product is defective if its safety is not such as a person is entitled to expect, taking into account all the circumstances including the presentation of the product, the use to which it could reasonably be expected to be put and the time when it was put into circulation.[5] This last factor is included to protect manufacturers from allegations of defects when product safety improves over time: the fact that a safer version is developed later does not mean that the earlier release becomes defective under the Directive. This is one potential defence which could apply to nanotechnology products in the light of the rapid evolution of both materials and technologies, but it will be essential to keep records as to the time when a particular batch is put into circulation or incorporate other means of identifying the date, in order to prove that the defence applies.

The same date is also relevant to the longstop defence under Article 11, which states that any person's right to claim damages are extinguished 10 years after the particular product which has injured them was put into circulation. In this context, where complex products are concerned there may be a number of possibilities as to

2. Article 3(3). This was interpreted by the ECJ in *O'Byrne v. Aventis Pasteur SA* (Case C-358/08) as a positive requirement to give the information; a mere denial of being the producer is not sufficient.
3. Article 3(1).
4. Article 9(b).
5. Article 6(2).

when the period begins to run, as a product is passed between members of a group of companies for manufacture, assembly and/or distribution. In a case concerning vaccines, the question arose as to which of the Aventis Pasteur companies had put the product into circulation since the claimant initially sued the company which distributed the product, and only learned which company had actually manufactured the vaccine after the 10-year period had elapsed. The European Court of Justice ruled that once the product is in the form in which it is offered to the public – that is, fully finished and packaged – then it is put into circulation.[6] Every stage before this may be treated as part of manufacturing. There is, thus, a risk that a company which does not have any responsibility within a group for manufacturing the product itself but does engage in packaging may be the one facing liability, since its actions will start the clock running as to putting the product into circulation. Contractual relations within corporate groups may need to reflect this.

The Aventis Pasteur case also clarified the strictness of the 10-year longstop: it is absolute. English procedural rules, which would have allowed the claimant to substitute the correct Aventis Pasteur company as defendant to the claim, were ruled incompatible with the Product Liability Directive.[7] The only exception to this may be if the distributor company is in effect controlled by the producer company, so that the act of distribution is, in reality, an act of the producer.[8] But the fact that the two companies are parent and subsidiary is not necessarily enough to establish this level of control; actual control in practice would need to be shown.[9]

The question of what standard of safety persons are entitled to expect is a value judgment including the balance between price, performance and any instructions as to how it is to be used (or warnings as to uses to which it should *not* be put). The issue is one of legitimate expectations. The actual expectations that any given user has about the product's safety are not relevant – some consumers may be exceptionally demanding as to what risks may be posed by a product while others may be exceptionally resigned.

Only a limited number of defences are available. Under Article 7, a producer is not liable if:

- it did not manufacture the product in the course of a business or put the product into circulation. This is essentially a defence where university or other non-commercial research products find their way into a consumer's hands;
- it is probable that the defect did not exist at the time when the product was put into circulation but came into being afterwards;
- the defect is due to compliance of the product with mandatory public regulations;
- the defect (in the case of a component) is due to the design of the product into which the component has been fitted, or instructions given by the manufacturer of that product; or

6. *OB v. Aventis Pasteur MSD and Aventis Pasteur SA* Case C-127/04 [2006] 1 WLR 1606.
7. *Aventis Pasteur SA v. OB* (Case C-358/08).
8. Case C-358/08 at para. 52.
9. *OB v. Aventis Pasteur SA* [2010] UKSC 23.

– the state of scientific and technical knowledge at the time when it put the product into circulation was not such as to enable the existence of the defect to be discovered.[10]

This last is known as the 'state of the art' or 'development risks' defence and is highly pertinent to products already on the market which incorporate or have been produced using nanotechnology. If in future it emerges that such existing products did in fact pose a risk which is not known today, the producers will not be liable up until the point where the risk becomes known. At that point liability will arise but only in respect of sales after that date. However, Member States were given the choice as to whether to include this defence when they implemented the Directive so it may not apply in all EU countries.

Where it does apply, the question of precisely whose state of knowledge is relevant has been discussed by the European Court of Justice,[11] which clarified that the state of knowledge of the industrial sector concerned is not the limit. Rather, it is the state of accessible knowledge in general. So research results which have not yet been published would not fall within the scope of knowledge, nor might results published only in an Indonesian university journal in Bahasa. Note, however, that once it is known how to discover the existence of a particular type of defect, it is not a defence for a producer to say that they themselves were not aware of it, or that it was not something they were in the habit of using: liability will accrue if the means to discover this kind of defect existed, whether they habitually used it or not.

Notably, the fact that some act or omission of a third party contributed to the damage as well as the defect is not a defence: the producer remains liable. However, where the damage is caused by the defect and some fault of the injured person then the producer's liability may be reduced or disallowed altogether.[12]

The European Commission's 2006 Report on the operation of the Product Liability Directive[13] observed that Member States' varying legal traditions have led to differences in interpretation, including the concept of a defect and the scope of the development risks defence. There were also suggestions that the Directive is simply an inappropriate way to deal with design and information defects in novel products. Nevertheless, the Commission elected not to propose any changes.

3.2.2 THE GENERAL PRODUCT SAFETY DIRECTIVE

3.2.2.1 Scope

The second major piece of product safety legislation is Directive 2001/95/EC of the European Parliament and the Council of 3 December 2001 on general product safety (the Product Safety Directive).[14] It applies to all products sold to consumers.

10. Article 7.
11. *European Commission v. UK* [1997] All ER (EC) 481.
12. Article 8.
13. 14 Sep. 2006 COM(2006) 496 final.
14. OJ L 11, 15 Jan. 2002.

This replaced a 1995 Directive on the same topic whichhad been criticized for not ensuring consumer safety sufficiently and lacking clarity in various aspects. The new Product Safety Directive aims to achieve more active market surveillance, more effective enforcement and a new regime for rapid action to remove dangerous products from the market once they have been identified.

This Directive applies to all products if either there are no European regulations governing the safety of a particular category of products, or if the sector-specific regulations are insufficient. If the sector-specific regulations are insufficient they nevertheless continue to apply but the Product Safety Directive applies in addition. It also applies independently of the Product Liability Directive. In practice all products apart from food and pharmaceuticals are subject to the Directive; the sectoral regulations for industries such as toys, cosmetics, electrical products and medical devices apply in addition to the Product Safety Directive. Specific guidance is available as to the overlapping areas.[15]

3.2.2.2 The Safety Requirement

The Directive imposes a general safety requirement[16] on any product put on the market for consumers or likely to be used by them, including all products that provide a service. Second-hand products that have antique value or those that need to be repaired are not subject to this requirement, but products not initially intended for consumers but which may foreseeably end up being used by them, for instance by transfer from professional to consumer markets, are covered. So too are products which may not be owned by consumers but will be used by them, such as equipments used in sports and fitness centres.

A safe product is one which used under normal or reasonably foreseeable conditions of use (including length of use and any requirements as to putting into service, installation and/or maintenance) poses no threat or only a reduced threat which is acceptable and consistent with the Directive's objective of maintaining a high level of protection for the health and safety of persons. Factors taken into account include: the product's characteristics, packaging, instructions; its effect on other products with which it is foreseeably likely to be used; its presentation and labelling, including warnings and disposal instructions; and the categories of consumer at risk when using it, such as children or older people.[17]

A product is deemed safe if it conforms to EU safety provisions or, in the absence of Europe-wide rules, if it complies with the specific national regulations of the Member State in which it is being marketed or sold. The product is also deemed safe if it complies with a European standard established according to the

15. Guidance Document on the relationship between the General Product Safety Directive and certain sector directives with provisions on product safety (November 2003); and Guidance Document on the relationship between the General Product Safety Directive and certain sector directives with provisions on product safety – Second Chapter (November 2005).
16. Article 3(1).
17. Article 2(b).

procedures in this Directive. There are no such standards relating to the use of nanomaterials. Other relevant European standards or Commission recommendations (setting out guidelines on the assessment of product safety) will be considered, as well as:

- the standards of the Member State in which the product is being marketed or sold;
- codes of good practice as regards health and safety;
- the current state of the art; and
- consumers' reasonable safety expectations.

3.2.2.3 Manufacturers' Obligations

Manufacturers must only put on the market products which comply with the general safety requirement. For this purpose, 'manufacturer' includes: anyone in the supply chain whose handling of the product may affect its safety; anyone who puts their own brand on a product so as to present themselves as the manufacturer; and, where manufacture takes place abroad, importers. The manufacturer must provide consumers with the information necessary in order to assess a product's inherent risks, particularly when this is not directly obvious.[18] To provide against the possibility that a product is found unsafe only after it has been marketed, manufacturers need to have systems to be notified of any risk, trace products on the market, keep their distributors informed of testing and monitoring activities, keep a register of complaints and ensure these are adequately investigated.[19] It may be necessary to send out information to consumers who already have the product, warning them of the newly identified risk and how to minimize it. If a product is verified to be unsafe then the manufacturer must recall it.

Distributors (whose activities are not liable to affect the product's safety) are also responsible for assisting with this process.[20] They must keep and make available any documents necessary for tracing the product's origin; pass on information about product risks and co-operate with the actions being taken by the manufacturer or any competent authorities

If the manufacturers or the distributors discover that a product is dangerous, they must immediately notify the competent authorities of all Member States in which they are marketing the product and, if necessary, co-operate with them. There is no obligation on a distributor first to inform the manufacturer or agree whether an identified defect amounts to a safety risk, if they are making a notification, although obviously it will also make sense to do this. There is the potential for difficulties within commercial relationships if in fact the parties disagree over a notification, and it would make sense to have a contractual requirement in place for a distributor to at least inform the manufacturer as soon as it receives a complaint or reaches the conclusion that a notification is required. Guidelines published by the

18. Article 5(1).
19. Article 5(1) paras 2–4.
20. Article 5(2).

Commission[21] (the Product Safety Guidelines) explain that 'immediately' means 'without delay' as soon as the relevant information has become available, and in any case within 10 calendar days even if investigations are continuing. This should allow for at least some consultation between the potentially affected parties, and it is important to maintain a communications channel in case such a rapid consultation is needed. However, if the risk is considered serious then the 10-day limit is shortened to 3 days which leaves little time for discussion of any kind. Where the situation is an emergency, the company should inform the competent authorities immediately by the fastest available means – picking up the telephone springs to mind.

The decision as to the point at which sufficient information has been gathered to indicate that a product is dangerous, a serious risk or an emergency remains one which the manufacturer or distributor has to assess for itself. A single complaint is unlikely to be sufficient unless extremely worrying by its very nature – for instance, that contact with the product has caused toxic shock to a user and may therefore carry a previously unsuspected risk of causing severe personal injury or even death. The Product Safety Guidelines set out a framework for carrying out this risk assessment but this has been replaced, at least for the national authorities' use, by a more sophisticated approach under Annex 5 of Decision 2010/15/EU.[22] This Annex includes categorizations of severity for the possible damage that may be suffered and a table setting out what severity of damage at what probability amounts to a serious risk. Since the national authorities are charged with monitoring the safety of products on the market, as explained below, it would make sense for manufacturers and distributors to apply the same methodology in their own internal assessments. A key requirement for managing potential liability for breach of the Product Safety Directive's requirements is to keep comprehensive records of the testing which is done, the result and the assessment which was carried out, in order to be able to justify the categorization which was applied. These records will then need to be kept in an accessible format for as long as the product remains on the market.

The requirement for the manufacturer to notify all competent authorities, potentially in all 27 Member States, is onerous, but there is no other system envisaged in the Directive. While as a matter of practicality it would seem sensible for the manufacturer to notify the authorities of the Member State in which it is based, and for that authority then to deal with the Commission and other Member States, this is not expressly permitted. A manufacturer taking this approach therefore risks being found not to have complied, unless the authorities of each Member State have announced that they are prepared to accept such relayed notifications as sufficient to meet the notification requirement. An alternative electronic process, the Business Application,[23] which was set up in 2009 enables producers and distributors to submit a single notification centrally, but this constrains the

21. Guidelines for the notification of dangerous consumer products to the competent authorities of the Member States by producers and distributors in accordance with Art. 5(3) of Directive 2001/95/EC.

22. OJ L 22/1, 26 Jan. 2010.

23. See <http://ec.europa.eu/consumers/safety/rapex/docs/manual_form_business_app.pdf>.

information which can be notified and makes the maintenance of confidentiality over any of the information submitted more difficult. The ability to keep information away from the public eye is limited since the Directive only stipulates that this can be done in respect of professional secrecy,[24] an undefined term which in some Member States is interpreted as being limited to technical trade secrets. It is therefore important to consider carefully what information needs to be included in any notification. Information on the risks presented by products must be provided to the public regardless of professional secrecy, if the circumstances so require.

Even if the notification can be done through a single channel, it will still be necessary for the manufacturer to liaise with the enforcement officials in each relevant market to manage the impact on their local distributors and customers rather than leaving this to the local authority. The best approach, therefore, is where possible to accept the administrative burden of notifying the national competent authorities individually and, using that process, to begin a co-operative approach to the problem.

3.2.2.4 RAPEX

The Commission manages a Rapid Information System called RAPEX to disseminate product safety information through all the Member States and also countries outside the EU, including in particular the United States and China. This international cooperation is being progressively extended, by the OECD and other international organizations which focus on consumer product safety. The Commission also publishes information about products notified to it as posing serious risks on its website.[25]

The objective is to enable the distribution of dangerous products to be limited or prevented on an emergency basis.

Once a product has been identified as posing a serious threat, the Commission can implement rapid measures at European level such as: adopting a decision imposing specific safety limitations; prohibiting the use of a substance; or requiring manufacturers to affix warnings to the product. These decisions last one year but can be extended by successive one-year periods.

3.3 CONTRACTUAL PRINCIPLES

3.3.1 Overview

Any party in the supply chain can be liable to their customer for a breach of any express or implied term in the contract, either through performing or failing to perform any obligation. The questions to consider are thus:

- What obligations are imposed through the contract?
- What will amount to performance or not?
- What remedies are available if the contract is found to have been breached?

24. Article 16(1).
25. <http://ec.europa.eu/consumers/dyna/rapex/rapex_archives_en.cfm>.

Contract law is an area where the laws of Europe are still not harmonized, although this must be qualified by pointing out that the laws of several continental European countries are closely related having all been harmonized in the nineteenth century when the Napoleonic Code was instituted. This means that similar principles apply in France, Spain, Italy, Portugal, the Netherlands, Poland and Romania; while the civil codes of Austria and Germany were influenced by it. The basic principle of freedom of contract applies, but subject to numerous special provisions in the code such that a civil law contract must be read in conjunction with the applicable version of the code. There are in addition requirements of good faith as to the performance of contract which affect the interpretation in a way not familiar to common law lawyers.

The contract laws of England and Wales, Ireland, Malta and Cyprus on the other hand are based on common law principles, and these form the basis of the discussion below. Freedom of contract is a fundamental principle of the common law, stipulating that parties with legal capacity can legally agree on anything which is not prohibited by another area of law. The most noteworthy restriction on freedom of commercial contract is competition law's prohibition on agreements affecting fair competition in the market, discussed further below.

3.3.2 Supply Chain Contracts

In respect of contracts between parties in a supply chain for the ultimate manufacture and distribution of a nanotechnology-enabled product (i.e., one which either includes nanomaterials or is produced by a process relying on nanotechnology, either materials, devices or techniques), contractual issues are most likely to arise around questions of definition and quality. For instance, is a customer entitled to reject a batch of a substance in nanoform if only 80% of particles have dimensions less than 300 nm, and only 70% below 100 nm? If the contract did not stipulate any particular particle size distribution, then it is not clear that this would be a breach of contract. The issue is even less clear if the product was ordered by reference to the purpose for which it was required, since a batch with a minority of larger particles may still perform the function but may perhaps result in a lower quality composite, or require different handling processes. Whether or not this amounts to a breach will depend upon the circumstances of the contract: did the supplier know the function for which the batch was ordered and the specification necessary for the manufacture of the composite? Is the reduced quality so poor as to be useless, or only saleable for a slightly lower price?

The solution in contract law terms is to avoid such arguments by stipulating both the characteristics of the product to be supplied[26] *and* the function it is intended to perform in detail in the contract. Although a contract does not need

26. And where nanomaterials and structures are concerned, it may be as well also to stipulate how the property is to be measured to determine compliance, since different methods may give different results.

to be in writing in order for it to be enforceable in English law, in practice a written and signed contract is the best form of evidence as to what was agreed.

The fact that the parties have taken the trouble to specify characteristics of the goods and their purpose in detail is also valuable in determining what remedies are available if the contract is alleged to have been breached.

3.3.3 Restrictions on Freedom of Contract

Cross-border supply agreements which include restrictive IP licensing terms, or tie the purchase of one product to the purchase of others which could otherwise have been sourced elsewhere, may infringe Article 101 Treaty on the Functioning of the EU[27] or equivalent provisions under national laws.

There have been many cases over the years to establish which clauses in a contract may or may not restrict competition, or be justified as objectively necessary, and this is an area in which specialist advice should always be taken.

Competition law also prohibits the holder of a dominant position in a particular market from acting abusively.[28] Ownership of a patent may put a company into a dominant position if the effect of the patent is that no other company can manufacture a product for which there is no direct substitute. This does not, however, mean that a refusal to grant a licence (for instance, to enable a customer to have a second source of supply) is automatically an abuse of that dominance: the right to maintain a monopoly is of the essence of a patent.[29] Nor is fixing a price higher than that a totally independent manufacturer would charge automatically an abuse, since the originator of a product is entitled to a higher rate of return to compensate for the costs incurred in perfecting it.[30] But there may be 'exceptional circumstances' in which the exercise of patent rights might constitute the abuse of a dominant position. The three conditions necessary for this are that:

- a new product, for which there was consumer demand, would be prevented from being brought to the market because the subject matter of the patent in question is essential to that new product;
- there is no objective justification for the refusal to grant a licence; and
- effective competition on a market is effectively excluded by the patent holder's actions.[31]

In these circumstances, the competent authority (the European Commission or a national competition regulator as the case may be) may be able to force the patent

27. Discussed in more detail in the context of patent licensing at s. 2.5.4 above.
28. Article 102 Treaty of Lisbon.
29. *Volvo v. Erik Veng (UK) Ltd* Case 238/87 [1988] ECR 2605.
30. *CICRA v. Renault* Case 53/87 [1988] ECR 6039.
31. *Magill [1995] ECR* 1-743 as confirmed and extended in Case T-201/04, *Microsoft v. Commission*, [2007] ECR II-1491.

holder to grant a compulsory license. In practice no such licenses have been ordered to date: in reality few monopolies are both strong and valuable enough to warrant the cost of bringing such a claim.

3.4 TORTIOUS PRINCIPLES

The relevance of tort law to defective consumer products has been largely eroded by the implementation of the Consumer Products Directive and General Product Safety Directive, discussed above. These are however directed towards enabling consumers to bring claims rather than businesses – the former entirely, the latter primarily though not exclusively. Thus, they may not cover the situation where a product supplied by one business to another causes loss other than through causing damage to consumers. For instance, if a material supplied in the form of nanoparticles was used in constructing equipment for the manufacture of other products, or interacted with other constituents in a manufacturing process such as to cause a fire, then these Directives would be unlikely to apply. Although the regulatory system in the form of Registration, Evaluation, Authorisation and Restriction of Chemicals (REACH)[32] should be sufficient to ensure that only materials whose safety is known are able to be put onto the market in the EU, chemical safety does not equate to fitness for purpose or necessarily address the compatibility of a material with other materials in a manufacturing process. Situations may, therefore arise, which are not included within these various regimes.

Where issues arise between a buyer and seller of goods, contract law will almost always provide the most appropriate remedy. Where the product causes harm or loss to someone other than the buyer, a different approach is needed. Independently of the specific statutory regime for liability in respect of consumer products, European legal systems normally recognize a general duty of care and provide for liability where some form of fault by one party causes damage to another, irrespective of any contractual relationship between them. The precise handling of liability varies: German law-based systems normally place greater emphasis on the intention of the person causing the harm, whereas French law-based and common law systems are more ready to impose liability where harm occurs irrespective of the intent of the person committing the act. Indeed, in some circumstances the French courts are the most generous, inferring fault from the fact that harm has occurred.

The common law approach is something of a middle way. The basic principles of negligence apply: a person injured or caused loss by a product has to establish that the injury or loss resulted from a failure by someone who owed a duty of care towards the victim to meet the appropriate standard of care. The issues which arise are thus whether or not the person accused did owe any duty of care to the injured person, and what was the standard of care which they were obliged to take?

32. See Ch. 6.

The manufacturer of a product is liable to the end user insofar as it sells a product in the form in which it is expected to reach the end user.[33] This responsibility includes the duty to inspect components bought in from third parties, and to issue warnings as to any danger which is discovered about the product only after it has gone into circulation.[34] For this purpose, a business purchaser of nanomaterials which intends to incorporate the product into another product can be seen as the end user since the nanomaterial is no longer identifiable as a separate item on the market once it has been so used.[35] Indeed, if the manufacturer of the nanomaterials knows the product into which the material is to be incorporated, then it could be liable to users of that product if harm results from the nanomaterial failing to meet the specifications necessary for that use, on the general principle that a manufacturer is liable to anyone whom it can reasonably foresee may be injured by its acts.[36] Of course, if the product can safely be used in one way but not in another, then a manufacturer which fails to give instructions on safe use will be liable for any harm caused by unsafe use. But if there is a possibility that the problem arose from the way the materials were handled or incorporated, then this may provide a defence.[37]

This responsibility can also apply to the distributor of the product, if they could reasonably be expected to carry out an examination of the product. It seems unlikely, given the difficulty of detecting let alone characterizing nanomaterials, that such a duty would be imposed on distributors.

The standard of care which must be applied is that which a reasonable person, guided by normal considerations, would apply. It can be negligent either to do something which such a person would not do, or to omit to do something which they would do. This necessarily depends upon the state of knowledge at the time,[38] and the standard which companies dealing in nanotechnologies will be expected to apply will evolve rapidly as more scientific information becomes available as to the precise nature of any risks which may be associated with them, and the best means to avoid or minimize those risks. A company is not necessarily expected to be immediately aware of every paper that is published and instantly adapt its own practices and products accordingly, but information which is widely available and accepted will be expected to be taken into account within a reasonable time. What should be done depends upon the magnitude of the risk of injury, the extent to which any change would ameliorate or eliminate that risk, and the burden of the steps necessary to make the change. It is not necessary to expend a great degree of effort to make a change which will have only marginal impact, unless the risk is otherwise intolerable.[39] But if a change which can be made without excess

---

33. *Donoghue v. Stevenson* [1932] AC562.
34. *Hollis v. Dow Corning* (1996) 129 DLR (4th) 609.
35. *Vacwel Engineering Co Ltd v. BDH Chemicals Ltd* [1971] 1 QB 88.
36. *Donoghue v. Stevenson, supra* n. 33.
37. *Evans v. Triplex Safety Glass Co Ltd* [1936] 1 All ER 283.
38. *Roe v. Minister of Health* [1954] 2 QB 66.
39. *Alexis v. Newham LBC* [2009] ICR 1517.

disruption to business and within the normal process of maintenance of equipment is not made, then the reasonableness of that decision may be open to question.[40]

What is reasonable may also be assessed by reference to prevalent industry standards, if these apply to the product or process in question. Although at the date of writing it would hardly be possible to argue that the nanotechnology industry had reached any consensus over standards either as to material characterization or safety issues, work is progressing apace in various standards organizations.[41] If any such standards reach a level of general acceptance, then whether or not formally adopted or mandated by regulators, the question of whether or not a manufacturer complied with the standard may in time become relevant to the issue of liability for negligence. Similarly, the voluntary codes of conduct which have been formulated[42] may in time come to represent a minimum standard for the conduct of businesses involving nanomaterials, such that a failure to abide by their guidance could result in liability in negligence. To date, however, it seems unlikely that any of them have achieved sufficient standing to be taken as more than best practice.

40. *Haley v. London Electricity Board* [1965] AC 778.
41. See Ch. 2.
42. See Ch. 4.

Chapter 4
Regulatory Approaches

4.1 INTRODUCTION: THE ROLE AND LIMITATIONS OF REGULATION

Regulation is fundamental to modern government, and may serve a number of overlapping and potentially incompatible purposes. Regulations can guard against specific known hazards, such as the inspection of restaurants for compliance with food hygiene requirements. They may identify potential new harms, as in the premarket testing requirements for new pharmaceuticals and pesticides. They may be tailored to secure specific benefits or distribution of benefit, or protect against perceived moral hazards, as in the requirement for licensing of any research carried out on human embryos. Fundamentally, like any other form of law, they are imposed to maintain that balance between competing interests which reflects the key values of a society. Commercial interests may favour unlimited market growth and consumer choice, citing the benefits of increased employment and advantages of new products over old, while consumer interests may advocate high standards of product safety. Governments may want to limit the effect of commercial activities on public health or the environment but also to minimize the cost of enforcing those limits. Selecting the right balance entails deciding what value to place upon, for instance, greater consumer choice compared with the potential risks and costs from manufacturing, using and disposing of the increased range of products. It is ultimately a political decision – as is the initial decision as to the fundamental approach to regulation: should regulation be the norm for all spheres of activity, or only exceptional?

In the context of development of new technologies, this question becomes whether development ought to be allowed without constraints, or somehow centrally planned and supervised. Central planning is wholly alien to the modern Western concept of freedom of scientific (and technological) enquiry, and also to the market economy basis upon which most modern research takes place.

In addition, such an approach begs the further question as to what objectives should be set, and what costs are and are not acceptable to society as a whole: what interests are to be prioritized? The answer is likely to vary depending upon which culture, nation or region is asked. Thus, despite the calls made by various civil society organizations, there is no prospect of any global governance being instituted for nanotechnology research and development. The issues will necessarily be worked out on a national or regional level, and there is no guarantee that the same outcomes will be reached.

The time dimension is also important: whilst businesses and consumers can adapt their behaviour to almost any set of regulations, there is no benefit at all in regulations which continually shift, making compliance an ongoing process of enquiry and adaptation. Regulations should therefore ideally be drafted to cover the circumstances as they are expected to be for at least the medium term rather than purely to deal with possibly ephemeral concerns. This is obviously the more difficult where, as in the current position with respect to nanotechnology, the state of knowledge is continuously changing.

Absent a detailed understanding of the risks to be addressed, and of the efficacy of existing protective measures, regulation can only specify principles of behaviour and allocate responsibility. However, binding regulation may not be the most appropriate mechanism for such a purpose.

4.2 APPORTIONING RESPONSIBILITY FOR UNFORESEEABLE HARM

Although it is tempting to imagine that there should be a single correct answer as to how the costs of unforeseen risks should be allocated, in fact there are a number of different approaches which could be applied. If a system aims to allocate costs in terms of economic efficiency it may reach one arrangement, whereas a system which attempts to allocate costs fairly as between the various actors – industry, government and individuals, for instance – will reach another. A purely economic approach might suggest a distribution of cost designed to maximize deterrence, to ensure that the sum of the costs of accidents and the costs of prevention is minimized. This ignores, however, the Western social norm that suggests there should be a link between injured and injurer through a system of redress for particular wrongs. Although potentially less economically efficient, such an approach does satisfy the human desire for acknowledgement of responsibility for actions. A system which created a single agency with responsibility for preventing accidents, for instance, and which was liable to compensate anyone who was a victim of an accident, might efficiently incentivize the agency to take precautions against accidents but would leave the victim, although compensated, feeling they had been rendered anonymous in the context of what had actually happened: the person who or which actually caused the harm is taken out of the equation. Such a system would also have low economic efficiency, since instituting a responsible agency separate from the actors whose behaviour holds risk is going to entail costs of supervision and enforcement which are likely to be higher than those resulting from the actors themselves taking responsibility for their actions. Whether such additional costs are acceptable or not is essentially a political question.

Such a hypothetical system also ignores the simple, universal truth that whatever precautions may be put in place *accidents happen*. No system can guarantee that no harm will ever occur. Given that, the distasteful but necessary question follows, as to how much cost is it worth incurring to prevent what level of harm. The tortious test of negligence arguably starts from this point: what is the standard of care which someone who has a legal duty not to injure his neighbour must apply?

Typical conditions for allocating responsibility for an outcome to a particular entity under tortious principles include:

- the responsible actor is an intentional agent concerning the action;
- the action, resulting in the outcome, was voluntary;
- the actor knew, or could have known, the outcome;
- the action of the actor contributed causally to the outcome; and
- the action was in some way faulty, that is, the actor can be blamed for the contributory action.

Clearly, in the present state of knowledge of nanotechnology risks the third of these conditions may not be met in a preponderance of cases – leading to the prospect that if harm is later found to be caused by nanotechnology, no one will be held legally or morally responsible. But even where some harm is found to have been foreseeable, the standard of care actors are generally required to meet in common law countries is that of the reasonable person – and a reasonable person does not move heaven and earth to prevent an injury which is perceived as minor, or one which has a relatively low likelihood of coming about. There is a limit, even for major injuries with a relatively high likelihood.

Further, an approach to allocation of responsibility based upon tortious principles can only go so far. Tort can be viewed as a corrective form of justice, under which an individual has a legal duty to repair any losses caused by his conduct in violation of the injured party's right not to be injured. What tort laws do not seek to do is go beyond whatever circumstances applied before the wrong was done and the harm suffered. If the distribution of goods – resources, rights, power – was unjust in the first place, then even if this injustice in some way contributed to the harm occurring, a tortious approach does not seek to correct this, save to the limited extent it incentivizes a potential wrongdoer to reach the standard of a reasonable person in avoiding inflicting harm. This failure underlies calls for a different approach to be taken to nanotechnology than has been taken to earlier technological developments. A new approach could redistribute both power and benefits in respect of the whole range of new possibilities opened up by such research, in particular in view of the potential harms which may accrue at the same time, and which are most likely to be borne by those disadvantaged by the prevailing inequalities.

Likewise, as discussed in Chapter 3, product liability law includes a 'development risk' defence such that a manufacturer which would otherwise be strictly liable for any harm caused by a product is able to avoid any legal consequence. While this may be an acceptable situation for incremental developments of well-understood technologies, nanotechnology represents the first example since the industrial revolution itself of a situation where a whole range of materials and devices may reach the market across a range of industries with inherent unforeseen risks.[1] If such risks do cause harm, then the victims will suffer without any mechanism for compensation or redress – a form of collective abdication of responsibility. It should be possible to devise mechanisms which reward higher levels of investment in safety and sustainability research, rather than leaving them to be treated as a cost as traditional market economics does.

One possible approach to judge the acceptability of unknown risks is by comparing them to other risks or to certain standards. One could, for example, compare the risks posed by engineered nanomaterials with those from naturally occurring (or unintentionally manufactured) nanomaterials. Unless and until the risk from engineered nanomaterials exceeds that from the extant sources, it is arguably unnecessary to impose additional precautionary measures. The fallacy in this approach is to assume that because a naturally occurring risk may sometimes be unavoidable, it must be morally acceptable and precaution is unnecessary.

Another approach is to base risk acceptance on the informed consent of people who are exposed to the risks, after informing them about these risks. This has the advantage that it inevitably includes the issue of fairness: people are unlikely to consent to a risk if they feel that its imposition is unfair in comparison to the distribution of benefit. But this approach works only when it is possible to inform: the unknowns in nanotechnology are still too great. Further, the risks may affect a large number of people at once, in different places and different contexts. Once again the issue of cultural and personal values as to what is or is not worth doing, and different responses to categories of risk or benefit, will interfere with the possibility of meaningful and collective consent.

A partial response which the European Commission and various national governments have currently arrived at is to fund from public sources research into these issues, thus imposing upon manufacturers the knowledge which may defeat any liability defence without also imposing the cost of obtaining that knowledge. This is, however, an enormous and, more importantly, a long-term task. It is in the nature of problems such as eco-toxicity that their impact is unlikely to be detected when a small sample is released only into one location; there will be no confidence that problems are not building up until the interplay of several different potential exposure paths are known over a period sufficient for the potential interactions between the nanomaterial and both animate and inanimate components of the ecosystem to have had effect. In effect, the funding of risk research is an insurance policy against major short-term risks being taken; any more gradual

1. The biotechnology revolution took place mainly in the heavily regulated sector of pharmaceuticals, where extensive pre-market testing should identify any significant harms before the product reached the marketplace. In agriculture the issues have given rise to extensive debate, which is now being expanded to include nanotechnology risks.

cumulative or transformative effects will only become apparent after years or decades of exposure. But lack of evidence of harm is politically an unconvincing basis for any action to be taken in the interim. It is particularly so when the potential social benefits of nanotechnology for applications such as water filtering and medicine seem substantial. The opportunity cost of prohibiting nanotechnology development (with its potential for contributing to badly needed economic growth) pending the completion of studies over a decade or more timescale, when there is no certainty that serious (irremediable) harmful effects will necessarily emerge, carries substantial political weight.

4.3 CODES OF CONDUCT, REPORTING AND ACCREDITATION SCHEMES

4.3.1 Introduction: Informal Regulatory Approaches

Regulation is not only about government-imposed legal obligations. Codes, guidelines, advisory committees, standards, incentives, professional or trade associations, compulsory or voluntary accreditation schemes and information systems such as reporting arrangements are also regulatory tools. These 'soft' forms of regulation are particularly important where the subject matter to be regulated is rapidly changing, providing as they do a form of governance which can respond more easily to developments in understanding. They are also a form which depends upon consultation and collaboration, since a self-regulatory environment operates through the affected group either initiating or selecting codes to which to adhere. Such schemes may thereby achieve an inclusiveness and legitimacy which strict 'black letter' law may not have where there is little agreement as to the object or means for achieving it. Such relatively voluntary measures also avoid the potentially adversarial quality of government regulation, and because no legislative mechanisms are involved can be much more rapidly instigated, amended and (if circumstances change) removed. The flexibility of such mechanisms can help in finding the most workable balance between the competing values of different interest groups.

4.3.2 Voluntary Reporting Schemes

Of these alternatives, the option which initially appealed to several governments – including the United States, United Kingdom and Australia – in connection with nanotechnology was that of implementing a voluntary reporting scheme. Such a scheme would serve several purposes:

- increasing the information available as to the exposure potential and safety information, what risk management practices may be in place in respect of the products and what materials are already on the market or in

development. This information would assist in the framing (if necessary) of suitable regulation to fill any gaps identified in existing regulatory frameworks;
- identifying gaps in knowledge to form the focus for additional research; and also
- increasing the transparency around nanotechnology research defusing some of the public suspicion of the work being done and increasing the potential for public acceptance.

Unfortunately, however, industry proved very reluctant to participate. The UK Department of the Environment and Rural Affairs' scheme, launched in September 2006, received only 13 submissions in its two-year pilot phase, and appears to have been quietly abandoned amid criticism that the scheme was too complex and that a UK-only initiative might undermine the potential for a level playing field across the EU.[2]

Similarly, the Nanoscale Materials Stewardship Programme run by the US EPA between 2008 and 2009 received only 29 submissions for its basic (reporting) programme and only four companies indicated willingness to sign up to the intended more in-depth testing programme which had been hoped to develop more complete data on representative nanoscale materials. Few of those submissions that were received included any health or environmental data, and much of the information was claimed to be commercially confidential.[3]

Australia's 2007 call for information on nanomaterials received just seven responses.[4] And in all these schemes, there is an obvious risk that such information as was provided may not be representative – voluntary reports are most likely to come from the most responsible corporate citizens.

In view of the lack of success of these initiatives, mandatory reporting is under consideration in Canada but has not to date been introduced. France's proposal to introduce such a scheme has been notified to the European Commission and is awaiting a decision, as discussed further in Chapter 6.

In some sectors, certification or accreditation schemes have been an effective way of managing without regulation. These involve a process of verifying what systems, process or products are acceptable or best practice, and any given applicant for the scheme has to demonstrate that its implementation of these comply with given standards. Limitation of risk is an obvious candidate for inclusion in the requirements. Since, however, the basic knowledge as to the risks nanotechnology presents and how and where they are most likely to arise is incomplete, and few nanotechnology standards are as yet complete, all that can be certified is system compliance with general risk management principles to reassure consumers that

2. Note of the 15th meeting of the nanotechnologies stakeholders forum 22 Jan. 2010, at <http://archive.defra.gov.uk/environment/quality/nanotech/documents/100305-meeting-note.pdf>.
3. EPA Interim Report, January 2009 available at <www.epa.gov/oppt/nano/nmsp-interim-report-final.pdf>. See further, s. 7.2 below.
4. NICNAS Summary of the 2008 call for information on the use of nanomaterials, November 2010, at <http://nicnas.gov.au/Publications/Information_Sheets/General_Information_Sheets/NIS_Results_Call_for_Information_2008_Nov_2010_PDF.pdf>.

the organization takes all due care and attention towards the potential risks. Without a nanotechnology-specific remit this approach does not seem to offer any additional governance value and is unlikely to achieve public credibility.

4.3.3 Codes of Conduct

4.3.3.1 Background

In the face of the rising outcry over the potential risks which nanotechnology may engender, industry was quick to recognize the importance of being seen to have the public's wider interests in mind in its research as well as the fundamental profit motive. The European Commission likewise recognized that good governance could be a requirement where nanotechnology research and development was publicly funded. Accordingly, in 2007–2008 a number of codes of conduct and risk management frameworks were developed and proposed, by industry, the European Commission and jointly between industry and non-governmental organizations. All share the objective of ensuring that nanotechnology is, and is seen to be being, developed in a safe, ethical and transparent way.

Codes, like accreditation schemes, may suffer from a lack of precise standards, enforcement or independent assessment of effectiveness and in the worst case may be perceived purely as a public relations exercise by the industry concerned aiming to stave off the introduction of legally binding rules. The lack of transparency as to what is actually being done to improve standards as a result of adherence to such codes is a significant bar to their credibility as an alternative to enforceable regulation. They are, nevertheless, a step towards acknowledging responsibility for the unforeseen risks which may arise from the research and development being undertaken, and a commitment to act in a way that should limit such risks by identifying them at an early stage.

4.3.3.2 The European Code of Conduct for Responsible Nanoscience and Nanotechnologies Research (the EU Nano Code)

The EU Nano Code takes the form of a Recommendation[5] to the Member States as to the principles and guidelines they should follow when formulating and implementing national strategies for developing sustainable nanotechnology research. It is not binding, and has been formally adopted so far only by Romania. However, in response to the 2010 consultation over possible revisions to the EU Nano Code, nearly 40% of respondents, 89% of researchers and 67% of industry, stated that their organization was already applying the Nano Code, suggesting that it is having some impact in practice.[6]

5. C(2008) 424 final, 7 Feb. 2008.
6. Analysis of results from the Public Consultation at <http://ec.europa.eu/research/consultations/nano-code/results_en.pdf>.

The Recommendation also positions the EU Nano Code as an instrument to encourage dialogue at all levels from policy makers down to society at large in order to increase understanding and involvement by the general public in the development of new technologies. It may be having less impact in this regard since a survey carried out by the NanoCode project[7] in 2010 found that most Member State governments had not communicated the Nano Code nationally and respondents were aware of few planning to do so.[8]

The EU Nano Code consists of a series of seven general principles and three guidelines with the underlying objective of guiding research in nanoscience in a safe, ethical and effective framework, so as to support sustainable economic, social and environmental development. It encourages cooperation. Notably, it relates to all man-made nanoscale objects, whether intentionally engineered or produced as by-products.

The seven principles are expressed in the broadest terms, leaving complete freedom of choice as to how they should be implemented. They are that:

- nanoscience and nanotechnology research should be comprehensible to the public, conducted in the interest of the well-being of individuals and society;
- research activities should be safe, ethical and contribute to sustainable development, neither harming nor creating a biological, physical or moral threat to the biosphere, present or future;
- the research should be conducted in accordance with the precautionary principle;
- governance of such research should be open, allowing for the legitimate right of access to information and allowing participation of all stakeholders in decision-making;
- best scientific standards including Good Laboratory Practices[9] should apply;
- research should be managed to encourage maximum creativity, flexibility and planning for innovation and growth; and
- researchers and research organizations should remain accountable for the social, environmental and health impacts their work may impose.

This last principle, of accountability, has been the subject of some controversy among respondents to NanoCode's survey, with the possible implication of legal liability being felt to be in conflict with the principle of encouraging creativity. This suggests a lack of understanding of the legal framework for liability for new products, since a non-binding code cannot of itself create legal liability although it may indicate a standard of conduct which could be taken into account under existing legal causes of action (discussed in Chapter 3 above). The principal form of accountability referred to here is more organizational, moral and ethical accountability – the

7. Funded by the Commission to support the take up of the EU Nano Code.
8. NanoCode Newsletter May 2011.
9. Established under Directives 2004/9/EC and 2004/10/EC.

concept that organizations should have someone whose role is to ensure that they take into account the potential social, environmental and health impacts of their research, consider alternatives which would eliminate or reduce those risks, and be prepared to justify any necessary risks by reference to the anticipated benefits.

Similarly, the need to act in accordance with the precautionary principle was felt to be a hindrance to open-ended creative research.

The response from the NanoCode project is the proposed development in 2011 of an online tool to support understanding and implementation of the EU Code. It is intended to translate the broad principles set out in the EU Code into a number of clear criteria, and to provide users with a simple and straightforward means of measuring their performance against these criteria. It will offer guidance to aid improvement of performance.

NanoCode are also working to develop comprehensive guidance aimed at improving awareness and understanding of the EU Code, supporting communication and dialogue concerning it, providing examples of measures that can be implemented by users to aid compliance with its principles, and suggesting ways for improving performance.

The Guidelines included in the EU Code cover good governance of nanotechnology research including an exhortation to make all information available, accessible and understandable by lay people as well as scientists and to ensure proper peer review of results before these are widely disseminated, in order to ensure both clarity and a balanced presentation. All efforts should be made to ensure the highest standards of scientific integrity in order to avoid raising public distrust; protection for whistleblowers is recommended as part of this. Inclusivity in deciding the direction nanotechnology research should take is recommended, through foresight exercises and consultation with relevant ethics committee, although no criteria are mentioned as to what the ethical review should take into account. Since ethics is very much a matter of cultures, this may lead to widely divergent interpretations in different Member States. Standardization of terminology and the development of methods and tools for risk assessment and metrology at the nanoscale are also to be included in research funding. The priority for funding should be projects aiming to protect the public and the environment, consumers or workers, and to reduce or replace animal experimentation. Conversely, no funding should be given for research which could violate fundamental rights or ethical principles, such as harmful viruses or the non-therapeutic enhancement of the human body. Deliberate intrusion of nanoscale objects into the human body should be avoided until risk assessment studies on long-term safety are available.

The second guideline relates to precaution, such as the development of guidelines on the prevention of pathologies induced by nanoscale objects. Good practice in labelling and classification should be applied, including the preliminary requirement of developing suitable communications such as a warning symbol equivalent to the symbol warning people entering a laboratory of the presence of radioactive materials. Funding proposals should be required to include a risk assessment, and funding should be given to the monitoring of the potential impacts over time. Research into the potential toxicology and eco-toxicology of nanoparticles should be undertaken, and also into the ethical, social and legal aspect of the new areas of research. The convergence between information technologies and biotechnology and neuroscience are mentioned for particular attention.

The third and shortest guideline relates to dissemination of information about the EU Nano Code itself and other relevant legislation.

In response to the comments in the 2010 consultation a revision of the Recommendation was expected in 2011, and consultation exercises were planned for that autumn. A revised draft may become available in early 2012.

4.3.3.3 Code of Conduct for Responsible Nanotechnology[10] (the Responsible Nanocode)

A group of four UK organizations including the Royal Society and the Nanotechnology Industries Association (NIA) produced the Responsible Nanocode in 2007, following a workshop involving a number of participants from a range of industries. It was launched in 2008.

The Responsible Nanocode is principles based rather than standards based and aimed to reflect good practice for companies and other organizations involved in nanotechnologies. The code was intended to offer indicators of good practice to guide responsible behaviour. It aimed to be international in scope and applicable to all organizations involved in developing, manufacturing, retailing, disposal and recycling of products using nanotechnologies.

The resulting code consists of seven key principles, expressed at a level of practical implementation steps which reflect the industry-orientation of its developers. The principle equivalent to the EU Nano Code's accountability principle is specifically stated to be accountability of the Board or an appropriate senior executive or committee. The EU Nano Code's principle of safety has as its equivalent in the Responsible Nanocode a commitment to ensure high standards of occupational health and safety for workers handling nanomaterials and nano-enabled products. Organizations subscribing to the code should also consider occupational health and safety issues for workers at other stages of the product lifecycle. Broader safety issues are reflected in the principle that subscribers should carry out thorough risk assessments and minimize any potential public health, safety or environmental risks relating to their products using nanotechnologies. As to any risks arising at later stages, organizations under this code are required only to 'consider the public health, safety and environmental risks throughout the product lifecycle' without committing to any action to minimize or restore damage arising. In particular there is nothing about a precautionary approach or any restrictions or limitations on the scope of nanotechnology research that may be carried on. However, subscribers also commit to consider and contribute to addressing the wider social, environmental, health and ethical implications and impacts of their involvement with nanotechnologies. What form this contribution may make is left entirely open.

10. Available at <www.responsiblenanocode.org>.

The Responsible Nanocode does require its subscribers to proactively engage with and be responsive to its nanotechnology stakeholders, including business partners, whom subscribers should encourage also to adopt it. Subscribers also commit to be open and transparent about their involvement with and management of nanotechnologies and report regularly and clearly on how they implement the Responsible Nanocode.

Each principle is supported by a set of examples of good practice, providing examples on how compliance can be demonstrated. For instance, the principle of Board accountability might involve articulating how responsibility for nanotechnology is assigned within the organization, publishing the commitment to responsible management and establishing mechanisms whereby employers or external stakeholders can raise concerns with the Board on any issue relevant to the organization's involvement with nanotechnology. The examples of good practice in relation to worker health and safety includes a suggestion to disclose any breaches of safety guidelines to relevant authorities including the impact and any action taken in response. The commitment to public health, safety and environmental risks includes the example that products be marketed only after the safety of the nanotechnology-enabled elements has been substantiated, and the assumption that the risks associated with a material in bulk form are the same as those associated with the nanoform is expressly rejected as a default position.

The Responsible Nanocode thus represents a considerable amount of detailed work as to how organizations should manage their work with nanotechnology in practice, in contrast to the EU Code's focus on matters of general principle and governance. Despite this, a review carried out by the NanoCode project in 2010 observed that there did not appear to have been much implementation of the Responsible Nanocode in the first two years.[11]

4.3.3.4 Corporate Codes

The chemicals industry has taken the lead in developing industry-specific codes, with several of the major companies setting up their own. These range in the level of detail and objective from Bayer's three pages largely focused on occupational health and safety issues (suggesting the use of closed systems, protective clothing and respirators), to the 100 pages of the Nano Risk Framework developed by DuPont in association with the Environmental Defence Foundation.

The International Council of Chemical Associations' overarching Responsible Care Global Charter[12] applies to all activities of the chemicals industry and has associated guidelines on product management from the point of view of occupational health, environmental protection and product safety. Although it is not specific to nanotechnology, its basic principles do correspond to some extent to

11. Synthesis report on codes of conduct, voluntary measures and practices towards a responsible development of N&N, September 2010, at <www.nanocode.eu>.
12. Available at <www.icca-chem.org/Global/Initiatives/RC_GlobalCharter2006%5b1%5d.pdf?epslanguage=en>.

some of the principles of the EU Nano Code. For instance, the Responsible Care Charter's requirement for adopters to report openly on performance, achievements and shortcomings is similar to the EU Nano Code's requirement of openness in governance of nanotechnology. It also requires continuous improvement in environmental, health and safety knowledge and performance of products and processes over the lifecycle to avoid harm to people and the environment, correlating to a much less strict version of the EU Nano Code's requirement that research should be safe, ethical and contribute to sustainable development, neither harming nor creating a biological, physical or moral threat to the biosphere, present or future. Finally the Responsible Care Charter does indeed mention responsibility, although far from the EU Nano Code's standard of accountability: it stipulates provision of help and advice to foster the responsible management of chemicals. This does not amount to a commitment to responsible management as such, nor even to require it from suppliers and customers, but merely to make it easier rather than harder to achieve. However, companies implementing the Charter are expected to initiate risk-based and cost-effective management measures to prevent negative impacts on health or the environment, in accordance with the precautionary approach.

The BASF Nanotechnology Code is based on the Responsible Care Charter. Like Bayer's, its specific guidelines focus on occupational health and safety, but it is much stricter in terms of the precautionary principle: BASF commits to marketing products only if their safety and environmental impact can be guaranteed on the basis of all available scientific information and technology. This would in theory imply an impossible standard – no product's safety can be guaranteed, let alone one about the behaviour of which in biological systems or the environment so little is known as is the case for most nanomaterials – so qualifications must in fact be understood to apply. BASF perhaps intend to commit that they will not market a product which is known on the basis of current scientific information to have safety implications or a negative environmental impact. Nevertheless, the code was produced internally, is self-policed and commits to no specific standards. What exactly does the statement that 'We regard it as our duty to provide information about both the opportunities and the potential risks of nanotechnology' imply? – publishing all research results as soon as they have been generated, or issuing a public warning only once a given nanomaterial has been established to be toxic? The answer of course is somewhere in between, and where exactly will be a decision taken case by case.

The BASF code also acknowledges the need to keep potential alternative monitoring technologies in mind, and commits BASF to engagement with government in drawing up standards and legislation; BASF is in fact a participant at the OECD Working Groups.

4.3.4 Risk Management Systems

4.3.4.1 Introduction

A code of conduct is in one sense a specific example of the more general category of risk management systems: by professing and maintaining particular standards of conduct an organization should reduce the risks which might otherwise arise from lack of awareness or negligence in carrying out particular processes or making decisions. Codes tend however to be drawn at a high level of generality and are therefore open to a range of interpretations when it comes to how to operate them on a daily basis.

4.3.4.2 Basic Principles

Any risk can be broken down into two contributing factors: the size of the exposure – how widespread the perceived threat may be and the concentration at which individuals are likely to encounter it; and the biological response to that exposure. A hazard which may cause relatively little harm when encountered only rarely can become a major risk if in practice individuals are likely to be exposed to it at high levels over a lengthy period. Conversely, a hazard which has the potential to cause severe harm may pose an entirely tolerable risk if the likely exposure is both rare and, even when encountered, transient.

Risk management is classically seen as a four-stage process, beginning from the identification of a hazard or potential hazard. Once a hazard has been identified, then it needs to be characterized by determining what adverse effects it may have and how serious these may be. This will include dose-response assessment if possible, or qualitative measures as proxies. Added to the hazard characterization, an exposure evaluation provides information about the likelihood and extent of exposure (of people, animals or ecosystems) to the hazardous material. This will cover both the sources, distribution and concentration of the material and the routes by which the population or livestock may come in contact with it.

Finally, the risk is characterized in terms of the probability of occurrence of harmful events of given severities. The answers will clearly depend closely upon the assumptions and data built into the hazard characterization and exposure evaluation: if a substance is wrongly characterized as mostly harmless then it will be characterized as presenting a low risk of future serious harmful events. It is for this reason that when the available data are inadequate or non-conclusive, there is a temptation to opt for the worst-case hypothesis. This will in most cases result in an exaggeration of risk compared to the (unknown) real risk, but does at least avoid underestimation.

A methodical approach to risk management can take some of the subjectivity out of such decisions, at least so far as internal decision-making is concerned. Several organizations have therefore developed risk management approaches specifically directed to nanotechnology research and development. These are further from any form of publicly enforced regulation than is a code of conduct, but may help in establishing public trust. The European Commission's SCENIHR in fact concluded that existing risk assessment methodologies are suitable for assessment of nanotechnology risks in most circumstances.[13]

---

13. SCENIHR Opinion The appropriateness of the risk assessment methodology in accordance with the Technical Guidance Documents for new and existing substances for assessing the risks of nanomaterials 2007.

4.3.4.3 NanoRisk Framework

The DuPont/Environmental Defence Foundation's NanoRisk Framework[14] has a rather different background from the corporate codes of conduct, having been produced not only through an unusual collaboration between industry and civil society but also being the product of subsequent consultation. In this respect, it is somewhat akin to the Responsible Nanocode. However, unlike that system the focus of the NanoRisk Framework is on the process for evaluating and addressing potential risks of nanoscale materials, and it includes copious detail on each stage of the process:

- describe the material and its application;
- develop profiles of the material's properties, inherent hazards and exposures at three stages of its lifecycle;
- evaluate the nature, magnitude and probability of the risks;
- assess the available risk management options;
- decide whether to continue development, document the decision and its rationale, inform stakeholders; and
- review and adapt the risk evaluation and consequential actions regularly.

This systematic approach aims to assist organizations in making sound risk evaluations and management decisions. It accepts head-on the fact that decisions will often be made on incomplete information and provides a mechanism to update assumptions, decisions and practices as more information becomes available. It also acknowledges the areas of concern which it does not address, such as social equity and personal privacy, suggesting that organizations which wish to can incorporate these elements into their own adaptations of the Framework. It addresses the accountability principle indirectly through stipulating that a specific person must be assigned the responsibility for ensuring the proper implementation of the Framework, and suggests the product steward for this role. It also touches on the question of transparency, though leaving open the crucial question of where the line should be drawn between complete disclosure and maintaining business confidentiality.

Like the OECD Sponsorship Programme, the objective of the NanoRisk Framework – and the primary differentiator from established generic risk assessment methodologies – is to develop profiles of each nanomaterial as to its properties, hazards and exposures. But the Framework goes well beyond the OECD project in addressing these over the entire lifecycle of the material. Where there are information gaps the Framework suggests taking 'reasonable worst-case assumptions', thereby embedding a precautionary approach, although the alternative option of using comparisons to other better characterized materials or processes may undermine this if applied before the point where it is possible to classify nanomaterials and technologies into groups of genuine comparables.

The NanoRisk Framework is a comprehensive approach the application of which may amount to a considerable burden for smaller companies. However,

14. Available at <www.nanoriskframework.com>.

although it is of course open to tailoring for a specific adopter's circumstances, the objective of the detailed and thorough approach is to achieve a high standard of understanding of the materials and associated risks. Consequently, cutting corners in order to reduce the time and cost of assessment necessarily implies a risk that information will not be generated which may have had an effect on the risk management approach which is adopted.

4.3.4.4 The Precautionary Matrix for Synthetic Nanomaterials

A similar initiative is the Swiss government's Precautionary Matrix for Synthetic Nanomaterials (the Swiss matrix at Appendix II), which is intended to assist private industry in complying with its obligations under existing legislation and in particular to apply the precautionary approach to the risk of release of nanoparticles or nanorods. It consists of an assessment method, helpfully spelled out in detailed accompanying Guidelines,[15] which will lead to the recommendation either to take precautionary measures over and above those generally applicable for working with bulk materials, or not. It does not include recommendations as to what precautions may need to be taken; being context-dependent, that is left to the user. The process of completing the preliminary matrix for a given material in a given context will identify knowledge gaps; a finished matrix should include the specific answers to the knowledge gaps which were identified. Notably, whereas the EU Code defines the nanoscale as 1–100 nm the Swiss matrix includes materials up to 500 nm, in order to provide a safety margin as far as aggregates and uncertainties in particle size distribution are concerned.

This model comes without any of the moral overtones concerned in accountability or transparency; it is solely concerned with understanding and managing risk on a technical level. However, in practice the more social issues are unlikely to be avoidable and implementers choosing the Swiss model may find that communication and accountability also need to be addressed, whether through existing corporate structures or by instituting new ones.

4.4 REGULATORY APPROACHES IN EUROPE

A series of past regulatory failures has led to a change in approach to regulation in Europe in recent decades. The EU has moved from a relatively non-interventionist approach to a more precautionary approach aiming actively to prevent damage to the environment and public health rather than wait to see whether any such damage ensues.[16] The failure of individual Member States or the EU to respond in a timely manner to the bovine spongiform encephalitis (BSE or mad cow disease) outbreak of

15. Guidelines on the Precautionary Matrix for Synthetic Nanomaterials BAG/BFU 2011, version 2.1, July 2011.
16. D. Vogel, *The new politics of risk regulation in Europe*, 2001, at <www2.lse.ac.uk/researchAndExpertise/units/CARR/pdf/DPs/Disspaper3.pdf>.

the 1980s, led to a major shift in public trust in government's ability to address food safety generally and, following the disagreements over the introduction of genetically modified organisms (GMOs) into food, new technologies. The public in many EU countries came to suspect that its governments prioritize economic interests over those most directly concerning individuals, namely their own heath and safety.

Other failures, such as the thalidomide episode of the 1960s where children were born permanently disabled as a result of a drug approved for the reduction of morning sickness amongst pregnant women, and the continuing impact on workers' health of asbestos, widely used for many years in the construction industry and only banned decades later after being belatedly recognized as highly toxic to those handling it regularly, have also dented public confidence in regulatory oversight.

As these examples demonstrate, governments can be fatally slow to act when faced with a difficult balance of unquantified risk and pressure from industry to leave things as they are. An increasingly informed public recognizes that it will have to provide the impetus for forcing action. The European Parliament has responded to public pressure by repeatedly calling for reviews of the regulatory position as it affects nanotechnologies in Europe, and rejecting the European Commission's conclusions from its 2009 Gap Analysis of the applicability of existing regulations to nanotechnology[17] that improved implementation of existing laws is sufficient. The European Parliament 2009 Resolution on nanomaterials:[18]

> 2. Deplores the absence of a proper evaluation of the de facto application of the general provisions of Community law in the light of the actual nature of nanomaterials;
>
> 3. Does not agree, before an appropriate evaluation of current Community legislation, and in the absence of any nano-specific provisions therein, with the Commission's conclusions that a) current legislation covers in principle the relevant risks relating to nanomaterials, and b) that the protection of health, safety and the environment needs mostly be enhanced by improving implementation of current legislation, when due to the lack of appropriate data and methods to assess the risks relating to nanomaterials it is effectively unable to address their risks;
>
> 4. Considers that the concept of the 'safe, responsible and integrated approach' to nanotechnologies advocated by the EU is jeopardised by the lack of information on the use and on the safety of nanomaterials that are already on the market, particularly in sensitive applications with direct exposure of consumers;
>
> 5. Calls on the Commission to review all relevant legislation within two years to ensure safety for all applications of nanomaterials in products with potential health, environmental or safety impacts over their life cycle, and to ensure that legislative provisions and instruments of implementation reflect

17. *Regulatory Aspects of Nanomaterials* 17 Jun. 2008 COM(2008) 366 final, with annexed Staff Document.
18. European Parliament resolution of 24 Apr. 2009 on regulatory aspects of nanomaterials (2008/2208(INI)).

the particular features of nanomaterials to which workers, consumers and/or the environment may be exposed;

6. Stresses that such review is not only necessary to adequately protect human health and the environment, but also to provide certainty and predictability to economic operators as well as public confidence;

14. Underlines the importance for the Commission and/or Member States to ensure full compliance with, and enforcement of, the principles of Community legislation on the health and safety of workers when dealing with nanomaterials, including adequate training for health and safety specialists, to prevent potentially harmful exposure to nanomaterials;

15. Calls specifically on the Commission to evaluate the need to review worker protection legislation concerning *inter alia*:
- the use of nanomaterials only in closed systems or in other ways that exclude exposure of workers as long as it is not possible to reliably detect and control exposure,
- a clear assignment of liability to producers and employers arising from the use of nanomaterials,
- whether all exposure routes (inhalation, dermal and other) are addressed;

23. Asks the Commission and the Member States to launch an EU-wide public debate on nanotechnologies and nanomaterials and on the regulatory aspects of nanomaterials.

Nevertheless, in the face of the unknowns surrounding the potential impacts of nanotechnology the scope for effective regulation remains limited.

4.5 THE PRECAUTIONARY PRINCIPLE IN LAW AND REGULATION

4.5.1 Emergence

In the past, an unknown risk would have weighed relatively lightly in the balance against the promise of technological progress and the benefit of new products and materials: the question of regulation would effectively have been brushed aside pending evidence of a demonstrable risk. But towards the end of the twentieth century the impact of human activity on the environment reached the stage beyond which the environment would suffer permanent damage. The initial response of regulators was to impose a 'polluter pays' principle, which subsequently evolved into a principle that action could and should be taken to prevent damage – a 'prevention principle' based upon the assumption that science could reliably assess and predict the results of human activities.[19] The emergence of ever less

19. *The Precautionary Principle*, report of the World Commission on the Ethics of Scientific Knowledge and Technology (COMEST) UNESCO March 2005.

predictable risks and the lack of confidence in science's ability to predict and/or remedy their consequences, together with the loss of public trust, has however pushed regulators towards a more precautionary approach.

Despite its relatively recent emergence, at a very abstract level the precautionary principle seems to be reaching some level of acceptance as the appropriate way forward. However, there is ample room for interpretation as to what (if any) form of regulation or action the principle actually requires. The principle was first incorporated in international law in the London Declaration on the Protection of the North Sea in 1987 and at a global level it appeared in the Rio Declaration of the United Nations following the Earth Summit in 1992.[20] Both of these used optional language however: definitions which permit but do not mandate action before a causal link has been established by absolutely clear scientific evidence. For instance, the Rio Declaration merely states that:

> Where there are threats of serious or irreversible damage, lack of full scientific certainty shall not be used as a reason for postponing cost-effective measures to prevent environmental degradation.

This is a relatively weak form of the principle, in that although prohibiting failure to act merely because the scientific evidence of harm is not yet conclusive, it does not actually require any intervention and certainly does not require any rigorous intervention if the costs are considered excessive. Nor do these formulations of the principle seek to require signatories to eliminate all unquantifiable risks – the London Declaration refers to 'the most dangerous substances' and the Rio Declaration to 'serious or irreversible damage'. What it does achieve, though, is to acknowledge that action can be taken provided that there is *some* scientific evidence of potential harm: the principle cannot be invoked merely in response to paranoid fears with no basis in reality. By accepting that action can be taken once some threshold level of scientific evidence suggests that precaution is needed, the principle places at least some weight on less than entirely proven threats when balanced against opportunity costs or even financial costs which may be incurred in taking preventative steps.

The 1992 Treaty on the EU (the Maastricht Treaty) made precaution for the first time a guiding principle of EU environmental policy. Article 130r(2) states that:

> 2. Union policy on the environment shall aim at a high level of protection taking into account the diversity of situations in the various regions of the Union. It shall be based on the precautionary principle and on the principles that preventive action should be taken, that environmental damage should as a priority be rectified at source and that the polluter should pay.[21]

What this precautionary principle consists of is not defined in the Maastricht Treaty, which prescribes it only in the one instance given above – to protect the environment.

---

20. Rio Declaration on Environment and Development, at <www.unep.org/Documents.Multilingual/Default.asp?documentid=78&articleid=1163>.
21. Now incorporated at Art. 191(2) of the consolidated version of the Treaty on the Functioning of the EU, OJ 2010/C 83/01 30 Mar. 2010.

But in the Commission's view the principle's applicability is much wider. This is supported by a Council Resolution of 13 April 1999 urging the Commission to be in future ever more determined to be guided by the precautionary principle in preparing proposals for legislation and in its other consumer-related activities and develop as priority clear and effective guidelines for the application of this principle.

4.5.2 Interpretation of the Precautionary Principle in the EU

In response, in February 2000, the Commission issued a Communication setting out its views on the precautionary principle.[22] It stated that:

> The issue of when and how to use the precautionary principle, both within the EU and internationally, is giving rise to much debate, and to mixed, and sometimes contradictory views. Thus, decision-makers are constantly faced with the dilemma of balancing the freedom and rights of individuals, industry and organisations with the need to reduce the risk of adverse effects to the environment, human, animal or plant health. Therefore, finding the correct balance so that the proportionate, non-discriminatory, transparent and coherent actions can be taken, requires a structured decision-making process with detailed scientific and other objective information.

The Communication's stated aims included building a common understanding of how to assess, appraise, manage and communicate risks that science is not yet able to evaluate fully, and also avoiding unwarranted recourse to the precautionary principle, as a disguised form of protectionism. The Commission also points out that reliance on a precautionary approach is not the same thing as a search for zero risk, which is rarely possible.

The Commission justifies the use of the precautionary principle by reference to international law – clearly mindful of the tensions already showing in 2000 in international trade relations over the EU's legislation on GMOs, discussed further at section 4.6 below. The justification is extended from various references in international environmental treaties and declarations to be claimed as 'a full-fledged and general principle of international law', although there is little in any non-environmental provision of the time to support this extension. There is, however, support in the case law of the WTO Dispute Panel Appellate Body, which has confirmed that there remains room for application of a precautionary principle in the interpretation of the WTO's Agreement on Sanitary and Phytosanitary Measures (SPS Agreement, relating to the application of food safety and animal and plant health measures with an effect on trade). In a decision relating to a European ban on the import of meat products from animals which had been treated with growth hormones,[23] the Appellate Body stated that:

22. COM(2000) 1, Communication from the Commission on the precautionary principle.
23. *EC measures concerning meat and meat products (Hormones)* AB-1997-4, at para. 124.

> Members have the right to establish their own level of sanitary protection, which level may be higher (i.e., more cautious) than that implied in existing international standards, guidelines and recommendations.

In particular, the Commission considers that the principle applies where preliminary objective scientific evaluation indicates that there are reasonable grounds for concern that potentially dangerous effects on the environment, human, animal or plant health may be inconsistent with the high level of protection which the Community has chosen to ensure. Scientific proof of cause-effect relationships are most difficult where the adverse effects do not emerge until after many years of exposure, so the potential long-term effects must be taken into account in considering risk management measures, since it may well not be possible by the time the hazard materializes to take action which will remedy the situation in the short term. Action at the time of exposure may, in fact, be proportional overall, even if the immediate risks appear low.

The recommended approach to risk has three components: risk assessment, but also risk management and risk communication. Only upon reaching an assessment – not an assumption – that there *is* a risk (even though it cannot yet be fully identified) does precaution in managing the risk become relevant. However, the extent of scientific knowledge must inform the approach taken to manage the perceived risk even though, by definition, the approach presumes that scientific evaluation is not yet able to determine the level of risk with certainty. Proper implementation of the precautionary approach requires not only as complete a scientific evaluation as possible but also some assessment of the extent of scientific uncertainty. Far from being a simple solution under which European regulators can simply ban products if they consider there is some risk, or take arbitrary decisions, the European precautionary approach requires a detailed grasp and objective approach to what is and is not known.

The Commission explicitly points out that in some cases the right political decision may be not to take any action, or only to commission further research. This decision is political rather than legal or scientific because it is a function of the risk level which is acceptable to the society upon which it is to be imposed. Even a decision not to take action should only follow where an assessment of the potential consequences of inaction has been considered. However, where action is to be taken on a precautionary basis the Commission directs that it should conform to the general principles of risk management. Those principles require that the action be: proportional to the chosen level of protection; non-discriminatory; consistent with any similar measures already taken; based on an assessment of costs and potential benefits of action or lack of action; open to review when new data emerges; and capable of assigning responsibility for development of the new scientific evidence necessary for a better risk assessment. The assessment of costs and benefits is not to be limited purely to economic costs but also to include issues of efficacy and the public acceptability of any measures under consideration. The comparison is between the most likely positive or negative consequence of the proposed action, and those of inaction, both for the short and the long term.

Notably, as to the assignment of responsibility for developing further scientific evidence, the Communication declines to make a general rule placing the burden of proof of safety of products in all cases on the manufacturer or importer. Such a system is one way of applying the precautionary principle but it is not the only way. If the product does not require pre-market authorization then it may be up to the public authorities, consumers' associations or users of the product, to demonstrate that it poses a risk.

The Commission also points out the range of actions available in response to precautionary principles. The alternatives are far from being either a total ban, or no regulation. Options include reduction of exposure, tightening of controls, adoption of provisional limits pending further evidence becoming available, recommendations for specific populations at risk, and so on.

This, then, is the approach which the Commission is attempting to apply in revising regulation to deal with the unidentified and unquantified risks which may be posed by nanomaterials. The principle is explicitly invoked in the existing and draft regulation of a number of sectors, discussed in Chapter 6 below.

4.5.3 The European Standard for Adopting Precautionary Measures

Internally, the legal regime of the EU sets a relatively low standard for the level of scientific justification needed in order for measures to be introduced to protect human health. For instance, in its 2002 decision in the cases brought challenging bans on the use of certain antibiotics in animal feedstuffs[24] the Court of First Instance[25] upheld the Council's 1998 decision[26] to ban the use of certain antibiotics as additives in animal feed. It concluded that despite uncertainty as to whether there is a link between the use of certain antibiotics as additives and increased resistance to those antibiotics in humans, the ban on the products was not a disproportionate measure given the need to protect public health. The antibiotics in question had been added in very low doses to the feedingstuffs of certain animals for many years as growth promoters, to reduce the time and food required for the animal to attain its required weight for slaughter. They were also said to prevent certain diseases in animals. However, scientists have claimed since the 1970s that the practice entails a risk of resistance to those antibiotics developing in animals and being transferred to humans, notably via the food chain, with the result that those and certain other related antibiotics could no longer be used effectively in human medicine. In 1997 and 1998, various specialist bodies including the World Health Organization (WHO) recommended the discontinuance of the practice.

24. Cases T-13/99 and T-70/99, *Pfizer Animal Health SA v. Council and Alpharma Inc. v. Council.*
25. Now known as the General Court.
26. Council Regulation 2821/98 of 17 Dec. 1998 withdrawing the authorization for the use as additives in animal feedingstuffs of virginiamycin, bacitracin zinc, spiramycin and tylosin phosphate.

The link between use of the antibiotics concerned and the development of resistance to those antibiotics in humans had not actually been proven in 1998, and so the Council invoked the precautionary principle in adopting the contested Regulation. Two major manufacturers brought proceedings to have the ban annulled, arguing that instead of undertaking a thorough assessment of the risks associated with those products, the Council and the Commission had attempted to exclude any risk, taking an unrealistic 'zero risk' approach, basing their decision on political expediency rather than on an objective scientific analysis.

The Court of First Instance set out the conditions on which the precautionary principle is to apply in Community law. It reaffirmed that in other cases dealing with food safety (in particular the 'mad cow' BSE crisis), the Community Courts had already held that it is possible to take preventive measures without having to wait until the full extent of the risks perceived become apparent. But a preventive measure cannot be founded on mere conjecture which has not been scientifically verified. It may be taken only where there is a real risk – that is, some positive probability that the negative effects which the measure is specifically designed to prevent will occur. The acceptable degree of risk cannot be set at zero. Accordingly, before taking any preventive measure, a public authority must carry out a risk assessment, which involves both a scientific and a political component. The first requires as thorough a scientific risk assessment as possible in the circumstances (including urgency). The view of the relevant scientific consultative committees must be obtained even if there is no legislative requirement to do so, unless there are equivalent scientific grounds for the proposed action. In this case, the Court considered that the authorities' knowledge of other antibiotics in the same class as one of the banned substances (bacitracin zinc) was sufficient basis for proceeding without taking an additional scientific opinion specific to that product. The second is a risk management exercise to assess what measure or measures are appropriate given the degree of risk set by it. The Court pointed out that the decision to ban a product is not a matter purely of science, but rather for the public authority to whom political responsibility has been entrusted.

On the facts of the case, the Court concluded that the Council's decision to ban virginiamycin, despite the competent scientific committee taking the view that there was insufficient scientific evidence to show a risk, was justified in the interest of protection of human health. The role of expert scientific advice is only to ensure that the measure in question is taken on an objective scientific basis. The public authority's responsibility is to meet the objective. Provided that the authority has properly examined all relevant aspects of the case carefully and impartially, including the reasoning of the scientific committee in support of its opinion, then it is not bound by that advice.

This case shows that the court will be deferential towards political and regulatory decisions taken on a precautionary basis even in the absence of clear scientific evidence, if the objective is sufficiently important to justify the measures taken. However, some evidence there must be, and whilst that is certainly the case with respect to some nanomaterials or classes of nanomaterials currently in use there may well not be sufficient evidence to go beyond the disapproved 'zero risk' approach for others.

4.6 REGULATORY DIVERGENCE AND INTERNATIONAL TRADE RULES

4.6.1 THE ANALOGY WITH GENETICALLY MODIFIED ORGANISMS

The issues facing regulators as they attempt to formulate an appropriate approach to regulating nanotechnologies bear a close parallel to those which faced regulators when genetic modification of organisms began to emerge in agriculture in the 1980s. As in the case of nanotechnology, the sheer novelty of genetic engineering technology meant that there was little knowledge and little scientific consensus around the question of what impact these new organisms might have either on the food consumers or on the environment within which they were produced. In that case, the different approaches to risk adopted by the European regulators and the North American regulators led to contested legal proceedings at the WTO.

4.6.2 THE WTO DISPUTE OVER FOOD DERIVED FROM GENETICALLY MODIFIED ORGANISMS

After much debate, the EU adopted a process-based approach so that the process of production of food derived from GMOs is the focus of suspicion (as to any impact on health and the environment from the modification of the plant) and hence of legislation. In contrast, the US approach is product-based and ignores the process provided that the actual product at market release is substantially equivalent to the conventional (non-genetically modified) form of the same product. This latter approach means that existing food safety regulation can apply, and any broader consequences of the production method are either addressed under other generic regulation (such as discharge of farm wastes) or not at all.

The EU began introducing specific regulation around the use of GMOs in 1990.[27] This provided a regime for the approval by the Council of Ministers of genetically modified crops on a case-by-case basis, and indeed the first such product – Monsanto's genetically modified soybean – was authorized in 1996. However, whether European trade associations saw in this an opportunity to protect their own market or European special interest and consumer groups suddenly woke up to the fact that genetically modified products were about to be introduced into the food chain, a backlash of public opposition rapidly arose and resulted in a number of European Member States introducing bans at national level – as permitted under the Deliberate Release Directive where there were justifiable reasons

27. Council Directive 90/220/EEC on the deliberate release of GMOs (the Deliberate Release Directive).

to consider that such crops posed a health or environmental risk.[28] Other Member States indicated that they intended to follow suit, unless a labelling regime was introduced which would enable consumers to be informed about the incorporation of genetically modified ingredients in food they were purchasing. Collectively, these objectors were sufficient in number to prevent the Council of Ministers from approving any further products. By 1998, therefore, the EU had in effect placed a moratorium on approval of further genetically modified crops for sale in the EU. Although the various national bans could have been subject to legal challenge by the European Commission on the basis that there was no objective justification, the issue was sufficiently politically divisive that no such actions were brought.

The impact on the agriculture of exporting countries which had approved a range of genetically modified crops for production, such as the United States, Canada and Argentina, was enormous, since the EU represents one of the largest importing markets for agricultural products in the world. The process-based nature of the EU requirements meant that there was no simple test which could be applied to the end product to demonstrate compliance; rather, farmers and exporters would be required to demonstrate that the products had been maintained segregated from genetically modified varieties of the same crop from planting through all processing stages. This implied handling and record-keeping procedures for which the American agricultural industry was simply not set up. Exports to the EU fell dramatically.

The United States, backed by Canada and Argentina, filed a complaint with the WTO in May 2003. In August 2003, as no resolution appeared possible the complaint became a formal request for a Dispute Panel to adjudicate the dispute.[29] The claims were that the effective moratorium on approvals breached WTO rules, as did also the various national bans on marketing even of EU-approved genetically modified crop products. The particular WTO rules alleged to have been breached were the SPS Agreement as well as the basic rules on trade included in the General Agreement on Trade and Tariffs (GATT). The SPS Agreement aims for harmonization of international health and safety standards in food except insofar as divergent measures can be scientifically justified[30] and are not applied in a discriminatory way. Scientific justification includes an assessment of the risks to human, animal or plant life or health. The specific issues in dispute were the provisions at Article 5.7 of the SPS Agreement, which states that:

> in cases where relevant scientific information is insufficient, a Member may provisionally adopt sanitary or phytosanitary measures on the basis of available pertinent information, including that from the relevant international organisations as well as from sanitary or phytosanitary measures supplied by other Members. In such circumstances, Members shall seek to obtain the additional

28. Deliberate Release Directive Art. 16.
29. *European Communities – measures affecting the approval and marketing of biotech products.*
30. Article 3.3. SPS Agreement.

> information necessary for a more objective assessment of risk and review the sanitary or phytosanitary measure accordingly within a reasonable period of time.

By Article 5.4, Members are also required to avoid arbitrary or unjustifiable distinctions as to what measures of protection are necessary. The question the Dispute Panel therefore had to answer was what level of scientific evidence is sufficient, and what period of time is reasonable for obtaining the additional information. The question of what amounts to an objective assessment is also implicit. Taken as a whole, the question before the Panel was whether a precautionary approach is compatible with WTO rules.

The European Commission evidently hoped that its proposed regulation on traceability and labelling of GMOs would help to defuse the dispute since it would overcome the objections of those Member States which were preventing further approvals. Accordingly, EU Regulation 1830/2003 was introduced[31] which requires a company selling genetically modified seed to inform purchasers of that fact (including what specific GMO is concerned) and the farmer must inform purchasers of the crop that it is genetically modified and keep a register of those to whom it has been sold. This is independent of the presence or absence in the actual product of detectable levels of modified DNA or protein. The limit above which labelling has to be applied for adventitious presence (the presence of traces not through deliberate inclusion but through external circumstances, such as crops being grown in a neighbouring field) of approved GMOs was lowered from 1% to 0.9%; for products pending approval it is 0.5%. This represented a tightening of the EU rules since under the previous regime, only products including detectable modified DNA or protein were regulated. Processing aids and enzymes are not regulated, nor are livestock products where the animal has been fed with genetically modified feed or treated with GMO-derived medicinal products.

The approvals requirements under the Deliberate Release Directive were also toughened, by including the need to consider indirect effects as part of any risk assessment as well as a post-marketing monitoring regime. On the positive side, a deadline for the approval process of 10 years from filing of the application was introduced.[32]

The United States proceeded to the WTO Dispute Panel despite these new regulations being introduced and the EU approvals process for GMOs re-starting. Indeed, far from accepting that this was an attempt to resolve the disagreement, some US officials viewed them as a further illegal barrier to trade; their status as legitimate food safety provisions was contested, and the additional burden of management, handling and record-keeping which US farmers would need to undertake in order to comply were said not to be practical. US farming organizations expressed concern that EU consumers would view the presence of labels as

31. Regulation on the traceability and labelling of GMOs and the traceability of food and feed products produced from GMOs OJ L 268/24, 20 Oct. 03.
32. Directive 2001/18/EC, OJ L 106, 17 Apr. 2001.

implying that a warning was being given about the product. The scientific justification for such measures was also argued to be insufficient.

The Panel upheld the US' objection to the national bans on EU-approved genetically modified products, and also upheld the charge that the EU approvals system did not operate without undue delay. The former ruling suggests that the scientific justification relied upon by the Member States which had implemented bans were found inadequate: the precautionary approach cannot justify an assumption of a risk where there is little or no evidence for one. However, notably the Panel did not accept the objections raised to the EU's risk assessment procedure as part of its approval process, including the requirement to take a precautionary stance, nor did it require any changes to the approval procedure itself. It made no recommendations as to how the EU should change its practices to come into line with WTO rules.[33] The ruling was not appealed.

This left entirely open the question of what, if anything, the EU was required to change. Settlement agreements were reached with Canada in 2009 and Argentina in 2010 under which the parties' respective competent authorities meet biannually to follow-up any authorizations of interest to a party, discuss measures relating to biotechnology which may affect trade between them and exchange information regarding proposed legislation. Diplomacy, rather than coercion, has been accepted as the way forward. The United States left the issue unresolved, facing an unpalatable choice between continuing dialogue with little expectation that the EU's labelling and approvals regimes are likely to change, or seeking retaliatory measures through the WTO on the grounds that the EU has not complied with the ruling. If non-compliance was established, this could pave the way for the United States to introduce higher tariffs on imports from the EU, but there is no mechanism which can actually force the EU to change its legislation.[34] Further, such tariffs are likely to entrench EU consumer attitudes in opposition to genetically modified foods through creating a perception that the United States is trying to force such products upon them contrary to its own governments' advice.

The same issues and tensions are likely to emerge over the EU's decision to introduce labelling requirements in respect of nanotechnologies used in cosmetics and novel foods.

4.7 ADAPTING REGULATION IN THE LIGHT OF FUTURE DEVELOPMENTS

The political pressures produced by the fear aroused by nanotechnology have resulted in an almost unique attempt to regulate the use of nanomaterials at the same time as they have emerged from pure research, rather than following their deployment in the marketplace. But as the repetitive and still far from conclusive

33. Dispute Panel Ruling WTO Doc. WT/DS291,292,293/R (adopted 21 Nov. 2006). At <www.wto.org/english/tratop_e/dispu_e/dispu_status_e.htm>.
34. The EU opted to accept tariff penalties in the earlier bovine growth hormone case, *supra* n. 23.

nature of the discussions being held between industry and the various regulatory and international bodies which are addressing the form of regulation show, the problem is intractable given the current state of knowledge of the materials and their potential behaviour across the whole range of biological and ecological contexts into which they could ultimately find their way through routes as yet unknown. Nevertheless, the same pressures dictate that *not* regulating nanomaterials is not an acceptable option. Whether through confirming the application of existing laws or by introducing new provisions specifically directed towards particular applications such as cosmetics or foods, regulators will continue to address nanomaterials as a category. Some of the current initiatives are discussed in Chapters 6 and 7 below.

Such regulations as are first introduced are likely to need to be re-evaluated as the state of knowledge of particular nanomaterials' properties increases. This should follow over time as reporting under REACH[35] and equivalent regulatory systems worldwide begins to deliver increasing levels of information about nanoforms of substances.

A further factor likely to influence future regulation is the shifting nature of the values which regulation seeks to balance. The perception of what constitutes an acceptable risk is not absolute and fixed, but relative to other perceived threats at any one time. The recessions of the early twenty-first century have already demonstrated that the strength of consumers' feeling about carbon-neutrality or eco-friendliness of the goods and services they consume is not immune to the impact on their spending power of increased debt repayment obligations and fear of unemployment. If nanomaterials have the potential to reduce the consumption of resources, improve the efficiency of renewable energy generation and reduce the cost of environmental clean-up from established sources of pollution, these benefits may well come to be valued more highly than they are at present, and the balance between unknown and possibly non-existent risks and concrete present benefits will shift in favour of greater permissiveness in the development and deployment of new materials.

Such a shift presupposes that the public is informed about both risks and benefits as the state of knowledge, and the economic and political circumstances, change. This requirement may increase the pressure for regulatory responses of the nature of labelling, since engaging consumers with managing their own exposure is a highly effective way of informing and keeping informed those who may be interested in influencing the debate.

The same shift may be manifest through the changing influence of different groups within society, as when the rise of the Green parties of northern Europe began to influence EU policy in the early 1980s and 1990s, thereby obtaining a practical influence in a geographical area far wider than their actual supporter base. The analogy of environmental regulation also points up the fundamental connection between regulation and the ability to monitor compliance and also the ongoing impacts of polluting activities. For regulation to be effective, governments need to

35. See Ch. 6.

be able to detect and prove such breaches. This means resourcing, both in terms of suitable equipment and also skilled officials able to deploy it appropriately. It is entirely likely that lack of suitable equipment and technical skills will affect monitoring of compliance with regulatory limits for nanotechnology. Developments in these areas may, in time, lead to stricter limits being imposed for exposure in various contexts (occupational health, environmental pollution).

Chapter 5
International Initiatives

5.1 INTRODUCTION

5.1.1 UN and Other Treaty-Based Organizations

Much discussion of nanotechnology policy issues is going on in international organizations, a number of which have a mandate to consider chemical safety within their respective individual areas of competence. The Inter-Organization Programme for the Sound Management of Chemicals (IOMC) was set up as an over-arching programme in 1995 to draw together and co-ordinate these various activities, covering the UN's Food and Agriculture Organization (FAO), Environment Programme (UNEP), Industrial Development Organization (UNIDO), Institute for Training and Research (UNITAR) as well as the Organisation for Economic Co-operation and Development (OECD), International Labour Organization (ILO) and World Health Organization (WHO). The UN Development Organization and World Bank are also observers on IOMC.

The activities and reports of the principal institutions of IOMC are discussed individually below, but these are not only international organizations taking an interest in this area. For example, the World Organisation for Animal Health (OIE), an international organization with 178 Member countries, has a mandate under the WTO SPS Agreement to safeguard world trade by publishing health standards for international trade in animals and animal products. Its *Manual of Diagnostic Tests and Vaccines for Terrestrial Animals* (*Terrestrial Manual*)[1] is intended to facilitate international trade in animals and animal products and to contribute to the improvement of animal health services world-wide. The objective is to provide internationally agreed diagnostic laboratory methods and requirements for the

1. Current (6th) edition at <www.oie.int/manual-of-diagnostic-tests-and-vaccines-for-terrestrial-animals/>.

production and control of vaccines and other biological products. Although the OIE has not yet proposed any instrument relating to the use of nanotechnology in animal feed or veterinary medicine, the most recent edition of the *Terrestrial Manual* includes for the first time reference to the potential applications of nanotechnology in diagnostics, which may be further developed as applications come into mainstream use.

5.1.2 International Risk Governance Council

The International Risk Governance Council (IRGC) is an independent non-governmental organization established in 2003 in the wake of large-scale international health problems such as Mad Cow Disease and Severe Acute Respiratory Syndrome (SARS) to help the understanding and management of emerging global risks that affect human health and safety, the environment and the economy and society at large. IRGC focuses on systemic risks for which governance deficits exist and aims to improve risk governance to maximize public trust in the processes and structures by which risk-related decisions are made. Many such risks, like those associated with nanotechnology, are complex, uncertain or even ambiguous, and the potential benefits and negative side-effects often interconnect.

The IRGC has done detailed reviews of the perception of risks associated with nanotechnology and possible regulatory responses. It has published non-binding risk governance frameworks for nanotechnology generally[2] and more specifically for nanotechnology in food and in cosmetics.[3] The former, a Policy Brief in 2007, categorized nanotechnologies into 'passive' (materials and particles which are simply incorporated into products for their effect on the material's properties) and 'active' (structures and devices which respond to their situation, either as sensors or through recognizing a release point for drug delivery and changing structure, for instance). At present, the former are far more prevalent in actual products than the latter, but this is expected to change as the technology is applied in more sophisticated ways. The latter are perceived to present potentially the greater risk, not least since they will give rise to novel products as opposed to the passive technologies' enablement of improvements to or evolution of existing products. The Policy Brief concluded that existing regulatory approaches are likely to be adequate to manage the risks of passive nanomaterials, albeit that thresholds and definitions may need to be amended to capture their pertinent characteristics. But the advent of active nanotechnologies may require new regulatory approaches.

Of equal importance to the practical aspects of managing specific manufacturing or use risks, the Policy Brief identified the difficulty of clear communication in the absence of agreed terminology as a potential cause of mistrust, since

---

2. IRGC Policy Brief: Nanotechnology Risk Governance Recommendations for a global, coordinated approach to the governance of potential risks 2007.
3. Risk Governance of Nanotechnology Applications in Food and Cosmetics, a report for IRGC prepared by Grobe, Renn and Jaeger of Dialogik GmbH 2008.

information may appear to be inconsistent or inaccurate depending on the choice of terms used. Public anxieties identified include: the potential for nanoscale devices to influence human behaviour through incorporation into the brain; the further erosion of privacy or independence through the wide deployment of computing devices; and the potential impact of nanoscale devices if released or escaped from human control. The authors identified the risk that the speculative nature of much of the discussion to date may further heighten the perception of untrustworthiness of information around nanotechology. Communication with the public and other stakeholders was proposed as part of the risk governance strategy, pointing out that benefits as well as risks need to be communicated since acceptance of any new technology is clearly linked to awareness of the potential benefits.

Interestingly, although published before the failure of the various national voluntary reporting schemes had become apparent, the Brief acknowledged that the voluntary risk governance systems so beloved of market-orientated thinkers often result in a lowest common denominator approach and may not deter those who prefer to operate outside such a system or not comply with it – whether because of competitive and investor pressures or merely a reluctance or genuine failure to recognize the potential for harm. It therefore identifies a need to consider anticipatory and coordinated measures where possible damage may be irreversible and significant – in other words, to apply the precautionary principle.

The 2008 paper specifically relating to governance of risk in food and cosmetics (the Governance Paper) aimed, in addition to the general principles laid out in the Policy Brief, to consider judgments as to the acceptability of nanomaterials in food and cosmetics made by a range of actors including different countries and international organizations, and to explore the possibility of a voluntary certification programme for labelling. Acceptability judgments relating to risk are influenced by personal and cultural factors: the former relate to one's familiarity with the threat, potential to control it and the risk-to-benefit ratio, while the latter include abstracts such as equity, vision as to future developments and the potential impact on one's interests and values. These latter factors lead to a noticeable divergence in attitudes to nanotechnology in food between consumers based in Europe and those based in North America. Europeans perceive the introduction of technology into food as a culturally unacceptable 'tampering with nature', whereas North Americans' concerns centre on the potential for using technology in food to harm consumers and exacerbate existing inequalities. The Europeans considered food and cosmetics particularly sensitive applications because of the direct contact between engineered nanomaterials and the body, when information from any source as to the potential risks from food is lacking and only limited information is available relating to cosmetics. The net result is a public perception that the industries promoting these technologies are behaving secretively. This led to mistrust given the close association consumers perceive between nanotechnology and incidents such as the Mad Cow Disease outbreak as well as the debates (or refusal to engage in debate) over genetically modified organisms in food. However, whereas this might lead one to expect that more information would result in less scepticism among consumers (since there is in reality no direct association between these subjects, and no long term harm has yet been shown to result from the use of genetic modification in food plants) in fact surveys discussed in the Governance Paper showed the opposite, with more informed consumers becoming more ambivalent.

As to voluntary codes, the Governance Paper reiterates the warning against the temptation to use voluntary schemes merely as 'window-dressing' to reassure consumer organizations without putting substantive self-regulation into place. Schemes established and operated in good faith can materially assist in offsetting consumer perception of poor risk governance. However, where codes proliferate, there is a risk that producers can claim the marketing benefits of being seen to be responsible but pick a code which in fact does not maintain the highest standards, for instance lacking enforcement mechanisms to compel proper compliance. The result may unfairly distort the market for nanotechnology products and ultimately once again lead to consumer mistrust.

Despite all such reservations, it may be unavoidable to have more than one code since one size cannot fit all participants in the nanotechnology supply chain around the globe.[4] It is also obvious that participation in the development of codes and standards, and adherence to such codes once agreed, is far preferable to the alternative strategy of maintaining secrecy over the accruing data relating to health and safety in the hope that sceptical attitudes will die down of their own accord. The IRGC cautions that past experience shows that although it may take many years, health consequences of foods and cosmetics will ultimately be identified and the company or companies responsible suffer massive loss of consumer trust (and brand value) as a result.

5.2 THE OECD

5.2.1 Chemical Safety Data

The OECD is an intergovernmental organization in which representatives of 31 industrialized countries in North America, Europe and the Asia and Pacific region, as well as the European Commission, meet to co-ordinate and harmonize policies, discuss issues of mutual concern and work together to respond to international problems. Observers from several countries with special status at the OECD, and from interested international organizations, attend many of the OECD's workshops and other meetings.

Although highly influential in many sectors, the OECD does not itself develop laws which bind its Member States. However, if Member countries consider it appropriate, an accord can be embodied in a formal OECD Council Act, which is agreed at the highest level of OECD, the Council.

There are two types of Council Act. A Council Decision is legally binding on OECD Member countries; whereas a Council Recommendation is a strong

4. Various nanotechnology codes which have been proposed are discussed in Ch. 4.

expression of political will but does not have legal force. In the area of chemicals, there is a Council Act relating to the Mutual Acceptance of Data (MAD).

The testing of chemicals is labour-intensive and expensive. Often the same chemical is being tested and assessed in several countries. In order to relieve some of this burden, in 1981 the OECD Council adopted a Decision[5] stating that data generated in a Member country in accordance with OECD Test Guidelines and Principles of Good Laboratory Practice (GLP)[6] shall be accepted in other Member countries for assessment purposes and other uses relating to the protection of human health and the environment.

Accordingly, although the OECD Act is not directly binding upon companies, where a product is required to be registered in more than one country the use of the OECD Test Guidelines and GLP save time and cost since safety data developed in one Member country must be accepted for use by the relevant registration authorities in for any safety assessment the chemical or product in another OECD country.

A further Council Act was adopted in 1989 to provide safeguards for assurance that the data is indeed developed in compliance with the Principles of GLP. This Council Decision-Recommendation on Compliance with GLP[7] establishes procedures for monitoring GLP compliance through government inspections and study audits as well as a framework for international liaison among monitoring and data-receiving authorities. A 1997 Council Decision on the Adherence of Non-Member countries to the Council Acts related to the Mutual Acceptance of Data in the Assessment of Chemicals[8] sets out a step-wise procedure for non-OECD countries with a significant chemical industry input to take part as full members in this system. Brazil, India, South Africa and Singapore participate at the time of writing.

5.2.2 Nanotechnology Safety

5.2.2.1 Initial Workshop

The OECD's engagement with nanotechnology safety was first addressed at the November 2004 meeting of the OECD Chemicals Committee, leading to a special session of the Committee on the potential implications of manufactured nanomaterials for human health and safety, which took place in June 2005. A workshop was held later that year to determine the 'state of the art' for the safety assessment of manufactured nanomaterials with a particular focus on identifying future needs for risk assessment within a regulatory context. Arising from the workshop it was

5. Decision of the Council concerning the Mutual Acceptance of Data in the Assessment of Chemicals 12 May 1981 – C(81)30/FINAL (Amended 26 Nov. 1997 – C(97)186/FINAL).
6. Available at <www.oecd.org/document/41/0,3746,en_2649_34365_1890473_1_1_1_1,00.html>.
7. Decision-Recommendation of the Council on Compliance with Principles of Good Laboratory Practice 2 October 1989 – C(89)87/FINAL (Amended 9 Mar. 1995 – C(95)8/FINAL).
8. 26 Nov. 1997 – C(97)114/FINAL.

recognized as essential to ensure the efficient assessment of manufactured nanomaterials so as to avoid adverse effects from their use. This led to the establishment of the Working Party on Manufactured Nanomaterials (WPMN) in September 2006, at about the same time as nanomaterials were first beginning to reach the market in commercial products.

5.2.2.2 Working Party on Manufactured Nanomaterials

The WPMN has met regularly since 2006 and has initiated a number of projects with the objective of gathering information and establishing testing and other methodologies to fill the gaps in existing knowledge of manufactured nanomaterials, and assess risks and the measures needed to manage them. The work is collaborative among the Member countries, with the lead (sponsorship) on any given project being taken by one or two countries and the outputs shared by all. Although not members of OECD, China, Thailand, South Africa, the Russian Federation and Singapore participate in this Sponsorship Programme.

An early project was the development of a global database of existing research projects and to enable additional research needs to be identified. The database was launched in April 2009, and will be extended as more data becomes available. It is hoped to lead to the creation of new collaborations and networks for researchers in the field.

The WPMN has also identified a representative set of manufactured nanomaterials with a view to testing their safety by reference to an agreed set of endpoints, without pre-defined regulatory consequences for any datasets. The materials selected as part of the representative set are: fullerenes; single-walled and multi-walled carbon nanotubes (MWCNTs); nanoparticles of iron, silver and gold; oxides of aluminium, cerium and zinc; dioxides of titanium and silicon; dendrimers; and nanoclays. The endpoints for this testing relate to: identification; physical-chemical properties and characterization; environmental behaviour such as biodegradability, accumulation or adsorption; toxicology both in the environment and in mammals; and material safety such as flammability. Issues identified by this initial round of testing are being further explored in a second round, and a Guidance Manual for the testing of Manufactured Nanomaterials has been published[9] which describes the testing programme for the specified set. The Guidance Manual will continue to be revised as more information is acquired. It is not expected that all endpoints included in the list will be suitable to test all of the manufactured nanomaterials. However, based on the practical results obtained from the first phase of the testing programme, it is expected that a clear indication will be provided on the appropriateness of the listed endpoints for those selected manufactured nanomaterials, as well as the need for further testing involving additional endpoints. The reporting of the results of Phase I of this programme began in 2011, and will provide an invaluable source of reference information for the understanding of nanomaterials in all applications.

9. ENV/JM/MONO(2009)20/REV, First revision 2 Jun. 2010.

The Guidance Manual will in turn serve as the basis for the development of comprehensive and consistent Dossier Development Plans (DDPs) for each form of manufactured nanomaterial, describing the testing programme including the identification of all participants, selection of the test materials and test designs to address all endpoints. A DDP will also review the existing state of understanding for each endpoint and set out the plans for completion of addressing each endpoint in view of that understanding. It is hoped that the process of developing both test methods and data will increase understanding of those properties of a manufactured nanomaterial that are unique to the nanoscale features of the material. For this reason testing is done against comparison non-nanomaterials, such as coarse materials, ionic material (e.g., ionic silver when compared to nanoscale silver particles), or impurities (such as nickel catalyst for single-walled CNTs). The process is thus necessarily iterative to some extent: as the properties become better understood, it will be possible to specify further information to be investigated in further testing. Eventually, the complete testing process should also take into account the life cycle of manufactured nanomaterials and evaluation of by-products of their use.

So far as appropriate, the DDP development is intended to use existing OECD Test Guidelines for materials, and accordingly a further project evaluated their adequacy for addressing manufactured nanomaterials. Although the project concluded that most existing test guidelines were adequate, this led to the publication in 2010 of Preliminary Guidance Notes on Sample Preparation and Dosimetry for safety testing of manufactured nanomaterials[10] since there is a significant risk of changes in the test material due to sample preparation and/or the nature of the test methods, that may affect the observed outcomes. The programme for developing DDPs is expected to be highly experimental and exploratory for this reason. The need to focus on dosimetry arises because of the widespread concern that existing regulatory thresholds for chemical substances, expressed in terms of the mass of a given chemical being put on the market, may be wholly inappropriate where the hazard potential of a material is due to its nanoscale properties rather than the mass in circulation.

Separately, the Working Party is facilitating co-operation to evaluate a suitable risk assessment approach for manufactured nanomaterials. A report on critical issues will be published in late 2011, setting out current practices and challenges and highlighting the research needed to enable adequate risk assessments to be made. A project on the use of alternative test methods in nanotoxicology, to reduce the use of animals in the hazard evaluation of nanomaterials, is at a preliminary stage, with initial consultations having been held only in April 2010 and a second expert consultation in January 2011. A case study for this project is in preparation.

The Working Party is also developing a report on the regulatory regimes, including voluntary schemes, being applied to manufactured nanoparticles by the OECD Member countries, to be published in late 2011.

Work on guidance for exposure measurement and mitigation for manufactured nanomaterials is also ongoing. In particular, the guidance on selection of skin protective equipment and respirators for workplace use and emission assessment for the identification of sources of airborne particles in the workplace are being updated.

10. ENV/JM/MONO(2010)25 31.5.10.

Finally, a project relating to the environmentally sustainable use of nanotechnology is in the process of preparing a report on national activities relating to the life-cycle assessment of nanotechnology. The interim version of this report was expected to be released during 2011.

5.2.2.3 Working Party on Nanotechnology

In addition to the WPMN, the OECD's Directorate for Science Technology and Industry has convened a separate Working Party on Nanotechnology (WPN). It was established in March 2007 to advise upon emerging policy issues of science, technology and innovation related to the responsible development of nanotechnology.

Since 2007, the WPN has been engaged in examining the business environment for nanotechnology, looking at a series of different industry sectors and also the over-arching policy issues. The aim is to identify and explore opportunities and challenges for nanotechnology commercialization and, where specific challenges are identified, to propose policy recommendations to address these. The key question is whether nanotechnology brings with it new and unique challenges for companies that may require new types of policy responses, including those in particular application fields.

In 2007–2008, the WPN undertook a project involving 51 company case studies across 17 countries. A forthcoming OECD report on challenges in the business environment draws on the company case study work and a survey of countries, both undertaken in 2008, and on previously existing reports and analyses. It identifies the origins and trends in the evolution of nanotechnology while also considering characteristics of the companies, the ways in which they are applying nanotechnology, and the challenges they are facing in commercialization. The case studies, supported by other data, indicate that challenges for companies arise across almost the whole scope of business activities, including research and development; human resources; finance; IP; value chains and production; public perception; and aspects of environment, health and safety. Of course, many of these challenges are faced by small companies in any high-technology field, but the essentially inter-disciplinary nature of many nanotechnology issues exacerbates them.

The first industrial sector being addressed by the WPN is Nanotechnology for Sustainable Energy. The project addresses the areas of energy harvesting (solar and wind) and energy storage (batteries and super-capacitors). A workshop in 2010 aimed to identify the challenges being faced specifically by companies working in the energy sector and using nanotechnology, and to discuss possible policy actions which could address the challenges. Issues identified included competitive pricing of renewable energy, the expense of raw materials, the complexity of technology transfer, industry awareness and the regulatory handling of nanomaterials. Of these, the first three are qualitatively similar to the challenges across the renewable energy and clean-tech industries.

The WPN has also carried out consultations in the area of business environments in nanomedicine to gather information which would help to further develop the proposal for work by WPN on nanomedicine.

A workshop on Nanotechnology and the Global Challenge of Access to Clean Water was held in Copenhagen in September 2008, to increase awareness and promote coordinated policy initiatives and fact-finding related to the potential for nanotechnology to contribute to the purification of water. Existing water purification techniques are relatively expensive, increase stress on watersheds and the environment, and are not readily transferable to the developing world. Emerging nanotechnologies such as nanofiltration and desalination, catalysts, new materials and sensors, may be able to contribute to improving this situation. However, the development, commercialization and diffusion of nanotechnology also faces many challenges, including the need to bring the laboratory work into the market without waiting for the perfect comprehensive solution to emerge.

In 2009 the OECD held a Policy Roundtable on Risk Governance Policy for Nanotechnology, to enable exchange of good practice in risk governance policy for nanotechnology through discussion between experts and delegates and hence support the development of risk governance policy. The Roundtable reviewed participatory models and the potential benefits of the involvement of a range of stakeholders in policy development for risk governance, as well as experience with voluntary measures by industry.

The WPN has also instituted a project looking at the potential environmental benefits of nanotechnology and the potential for economic growth through development and deployment of 'green' nanotechnologies, hosting a conference in 2009 on nanoscale innovation with the goal of encouraging development of technologies that can result in environmental gain (both cleaner production and reduction and remediation of existing pollution) while addressing unintended consequences. One factor which emerged from that conference was the global nature of the supply and distribution chains for nanotechnology products, wherever manufactured. As a result, a purely national or regional approach to regulation simply is not adequate to manage any environmental hazards which may in time be identified. Further, since engineered nanoparticles are extremely difficult to detect once incorporated into a product, a country which opted to ban their manufacture would nevertheless rapidly find products crossing its borders which infringed that ban and any attempt to enforce the ban across borders could lead to trade disputes in exactly the same way as environmental laws and labelling requirements have done in the past (as discussed in Chapter 4). If policy and regulation are to be effective, ultimately they will need to be substantially harmonized.

The WPN is also working on definitions of nanotechnology, in order to systematically develop indicators and statistics related to nanotechnology. It is also in discussions of nanotechnology metrics in the context of other emerging and converging technologies (e.g., biotechnology).

5.2.2.4 Publications: Safety of Manufactured Nanomaterials

The OECD Environment Directorate is publishing a series of monographs covering all aspects of the work being undertaken. In 2010 a report came out giving the results of an exercise to find out and compare what assessment criteria had been devised for the handling of nanomaterials in laboratories.[11] It compares existing published guidelines regarding the use of nanomaterials on a laboratory scale, including the manufacture and the use of products in industrial, institutional and commercial settings, since institutions as well as companies entering the nano-sector in all countries are producing and utilizing nanomaterials on that scale. As expected, a number of different strategies for assessment and implementation of protection measures have been developed in those areas. As might perhaps also have been anticipated, although there is a degree of consensus – as to use of the precautionary principle to minimize risk, for instance, and the use of vacuuming and wiping of surfaces by way of routine cleaning – the lack of large-scale data relating to exposure and risk meant that the guidelines reviewed embodied divergent opinions as to how occupational exposure to nanomaterials should be addressed. Some recommended that any new substances in laboratories, which are insufficiently examined for their properties including acute and chronic toxicity and physico-chemical characteristics, shall be treated as acutely toxic, caustic, chronically toxic, flammable, pyrophoric and explosive. Others support precautionary measures if the mass of the nanomaterial sample exceeds the milligram range but require nanomaterials below this mass to be regarded as potentially toxic if they are water insoluble and/or if the macroscopic material is classified as toxic. The majority of guidelines reviewed recommend the disposal of nanomaterials as hazardous or chemical waste.

The divergences continue with the issue of categorization, where at one end of the spectrum an assumption is made as to the equal potential toxicity of nanomaterials and at the other risk levels are to be defined by reference to the specific properties of the respective nanomaterial such as size, chemical composition and surface area. There is also a wide range of opinions as to what factors should be considered in a risk assessment, what respiratory protection is appropriate, what cleaning agents should be used and whether special storage arrangements for nanoparticles are needed.

In view of these incompatible opinions, it is clear that considerable work is needed before it will be possible to identify appropriate thresholds which can be used to draft regulations for all laboratories' or other organizations' handling of nanomaterials.

5.3 UN ENVIRONMENT PROGRAMME, ICCM AND SAICM

5.3.1 The UNEP

The UNEP is the United Nations' designated entity for addressing environmental issues at the global and regional level. Its mandate is to co-ordinate the development of environmental policy consensus by keeping the global environment under review and bringing emerging issues to the attention of governments and the

11. ENV/JM/MONO(2010)47.

international community for action. Its responsibilities include helping to develop international environmental law, which means that the development of new law is more central to its activities than it is to those of the other supranational entities discussed in this chapter. It also provides expert advice on the development and use of environmental concepts and instruments as well as promoting international co-operation in the field of the environment and recommending appropriate policies.

UNEP first looked at the environmental issues surrounding nanotechnology in 2007 in its report on *Emerging Challenges – nanotechnology and the environment*.[12] The report recognized the potential environmental benefits from exploiting nano-technology, as well as the potential risks. For instance, the use of nanomaterials to substitute for conventional materials may reduce the total raw material requirements and energy consumed in production for products across a whole range of sectors. Nanoparticles may enable better targeted delivery both of medicines and of plant protection products, reducing the impact on other species from the use of pesticides and herbicides. Nanofiltration has great potential for improving the availability of clean drinking water. Novel catalysts using nanoparticles are being developed to reduce the harmful emissions from a number of industrial processes and fuel additives for the same purpose are already on the market. Perhaps most ironically, given the dearth of information surrounding the prevalence and effects of nanoparticles themselves, nano-sensors are being developed which are able at very high sensitivity to detect and measure the presence of known pollutants in the environment, enabling more effective ways to be used to reduce their effects. For instance, a method for measuring concentrations of the pesticide atrazine in groundwater has been produced using nanoparticles of europium oxide.[13]

For all these potential benefits, however, the 2007 report equally acknowledges the concerns over the lack of knowledge of the behaviour of nanoparticles in the environment – how they may change over time, their impact on organisms and, by extension, ecosystems. Their toxicity, bioavailability, mobility, stability, solubility and reactivity require further study. The incredibly small size of engineered nanoparticles means that, like naturally occurring ultra-fine particles, they can remain airborne over very long distances, increasing their potential for cross-border impacts.

Finally the UNEP's Yearbook for 2010 maintained the focus on nanomaterials in a section on Harmful Substances and Hazardous Waste, reviewing and noting the various reports which have been issued on the lack of knowledge and potential risks involved in releasing nanomaterials into the environment. The Yearbook points out that the apparently benign use of nanoparticles to remedy contaminated groundwater and soil is already taking place, despite the lack of research into potential downsides. It appears that in practice, the precautionary principle is being applied with a light touch.

---

12. At <www.unep.org/yearbook/2007/PDF/7_Emerging_Challenges72dpi.pdf>.
13. Feng et al., *Functionalized Europium Oxide Nanoparticles Used as a Fluorescent Label in an Immunoassay for Atrazine*. Anal. Chem. 75 (2003): 5282–5286.

5.3.2 Strategic Approach to International Chemicals Management

The International Conference on Chemicals Management (ICCM), which operates under the aegis of the UN and the UNEP, now includes the issue of nanomaterials in its Strategic Approach to International Chemicals Management (SAICM), a policy framework set up in 2006 to foster the sound management of chemicals world-wide as part of work towards the sustainable development goals for 2020 set by the World Summit on Sustainable Development held in Johannesburg in 2002. SAICM is not legally binding in itself but does have significant high-level political support. It also includes a Quick Start Programme which provides guidance and funding for less developed countries to assist them in: the development or updating of national chemical profiles; the strengthening of national chemicals management institutions and programmes for sound chemicals management; and enabling the implementation of SAICM by integrating the sound management of chemicals in national strategies, and thereby informing development assistance co-operation priorities.

At its second (and latest) meeting in 2009 the ICCM adopted a Resolution[14] on nanotechnologies and manufactured nanomaterials which includes recommendations that Governments begin a public dialogue on nanotechnologies and manufactured nanomaterials and strengthen the capacity for such engagement by providing accessible information and channels of communication. The Resolution also recognizes the need for regulatory, voluntary and partnership approaches to promoting the responsible management of nanotechnologies throughout their life cycle and the need for development of these technologies to be consistent with the 2020 World Summit on Sustainable Development goals related to chemicals.

Since that meeting, various UN IOMC agencies have been working to assist Member countries to implement national chemical legislation and administration in accordance with SAICM. For instance, in 2011 UNITAR was beginning three pilot projects on nanotechnology in order to generate experiences and lessons learned that will be transmitted for deliberation at the third ICCM, to be held in mid-2012. It also ran a series of regional workshops over 2010–2011 to assist countries in developing nanotechnology policies and governance and provide guidance and training materials on topics such as the issues surrounding regulatory definitions of nanotechnology.

Looking ahead, there is a proposal to incorporate nanotechnology as an issue explicitly into the Global Plan of Action which is one of the three documents underpinning SAICM, at the third ICCM in 2012. A report on nanotechnologies and engineered nanomaterials, not yet available at the time of writing, is to be reviewed at that meeting. Delegates at the African regional meeting on SAICM in January 2011 called for this report to address: the critical role of the precautionary approach in addressing nanotechnology and manufactured nanomaterials throughout their life cycle; implementation of the 'no data, no market' principle prior to commercialization of nanotechnology applications; the need for producers to provide appropriate

14. SAICM/ICCM.2/15 Resolution II/4 E.

information about the content of manufactured nanomaterials to inform consumers about potential risks; and the ethical and social risks and benefits of nanomaterials, especially in developing countries and countries with economies in transition.

It remains to be seen what form of precautionary approach – whether permissive, like the Rio Declaration, or more active like the European Commission's interpretation – can be agreed by the delegates at the full ICCM meeting.

5.4 UN FOOD AND AGRICULTURE ORGANIZATION

5.4.1 The Role of the FAO

The FAO, founded in 1945, leads international efforts to raise levels of nutrition, improve agricultural productivity and contribute to the growth of the world economy. It is intended to serve as a neutral forum where nations meet to negotiate agreements and debate policy. The FAO also serves as a knowledge network on best practice in a number of areas. As part of this, it attempts to help developing countries and countries in transition modernize and improve agriculture, forestry and fisheries practices and ensure good nutrition for all. It also runs field projects throughout the world, providing technical know-how.

What the FAO does not do is instigate international treaties or other legal agreements which have any international force, of its own motion. It is a discussion forum and an instigator and disseminator of research into policy and technical issues which affect agriculture and fisheries in its Member countries. Where the Member countries' discussions identify a need for and consensus around an international approach to an issue, the FAO Conference may, under Article XIV of the FAO Constitution, '*approve and submit to Member Nations conventions or agreements concerning questions relating to food and agriculture . . .*' and consequently support the process of drafting an agreement. These tend to relate to establishment of various international commissions dealing with particular agricultural or fishery related issues, such as: the Plant Protection Convention relating to phytosanitary measures to control the international spread of plant pests and diseases, pest-specific measures such as control of the desert locus; or the establishment of thc International Rice Commission. It is unlikely to develop any legal measures relating to nanotechnology since this is a topic far broader than its remit.

5.4.2 International Food Standards: The Codex Alimentarius

The FAO is, however, alongside the WHO the co-sponsor of the Codex Alimentarius Commission,[15] which develops food standards, guidelines and related texts such as codes of practice under the Joint FAO/WHO Food Standards Programme. The resulting body of guidance is known as the Codex Alimentarius. The objective

15. <www.codexalimentarius.net>.

is to ensure fair trade practices in the food trade, as well as promoting coordination of all food standards work undertaken by international governmental and non-governmental organizations. In 1985 the UN General Assembly included in its Guidelines for Consumer Protection[16] the principle that when formulating national policies and plans with regard to food, Governments should take into account the need of all consumers for food security and should support and, as far as possible, adopt standards from the Codex Alimentarius. To this extent, therefore, the Codex standards are recognized as part of the body of normative though non-binding international law. In practice, products complying with Codex standards are permitted onto the market of the over 180 members of the Commission.

5.4.3 FAO/WHO Expert Meeting Report

The Codex standards do not currently address the issue of nanotechnology being used in food, but following on from calls by the International Union of Food Workers and Friends of the Earth for the use of nanomaterials in food and agriculture to be banned, and a ban being imposed on members by the UK's Soil Association, an expert meeting on the potential food safety implications of the application of nanotechnologies in the food and agriculture sectors was held in 2009 to discuss the use of nanotechnology in food production and processing, and the potential human health risks resulting from this use. The scale of the topic led the expert meeting to tackle the topic from a strategic level rather than that of detail and regulatory instruments. Occupational health and safety, and the environmental impacts, save insofar as the presence of nanotechnologies applied in the environment could affect the food chain, were specifically excluded from the discussion.

The resulting FAO Expert Report[17] discusses the current areas of application of nanotechnologies in food production, processing and packaging and the much larger area of potential uses still in the research and development phase. It observes that much of the currently available information related to potential applications which may not reach fruition for some time to come, and that actual applications of nanotechnology in food and agriculture are relatively limited for the time being.[18] The identification of products in circulation is also made more difficult by the fact that some manufactures avoid any reference to 'nano' in their communications as a result of the public concerns around these poorly-understood technologies – in contrast to the cosmetics industry where the high-technology connotations of using 'nano' in communications may have resulted in some products being marketed by that tag which in fact have no nanotechnology components at all.

Nevertheless, the FAO Expert Report found that the number of applications of nanotechnologies in the food industry is steadily increasing. The food industry is

16. A/RES/39/248; Guidelines at <www.un.org/esa/sustdev/publications/consumption_en.pdf>.
17. FAO/WHO Expert Meeting on the application of nanotechnologies in the food and agriculture sectors: potential food safety implications: Meeting report. Rome, 2009.
18. A comprehensive listing appears as Appendix 4 to the report.

constantly seeking new ways to offer products with improved tastes, textures, shelf lives and better safety and traceability. There is also ubiquitous pressure to reduce the level of salt, sugar, fat and artificial colours and preservatives being used in food. Nanotechnologies, such as ingredient delivery through nanoscale micelles,[19] and novel anti-microbial packaging materials better able to protect stored foods pending their consumption, have the potential to assist in reaching many of these objectives. Water-insoluble additives can be dispersed in food products as nanomaterials without requiring addition of other fats or surfactants. Nanoscaling may improve uptake and absorption of bioactive substances. Nanofiltration for the removal of undesirable components is also in contemplation, and the use of nanostructures to alter flavours and textures and improve consistency and stability of emulsions (enabling potential reductions in fat content without changing the consumer perception).

At the same time, however, past experience with novel additives has shown that unforeseen side-effects may arise, such as the alleged association between yellow food colouring and attention deficit hyperactivity disorder in children. There are therefore legitimate concerns as to whether the wholesale use of nanotechnologies to solve one set of problems may give rise to another. Whilst considerable research is going on to develop new nanotech-enabled products, relatively little is being done to evaluate whether these pose potential health or pollution risks if deployed on a large scale. Consequently, it is not possible to predict reliably the toxicological characteristics of engineered nanomaterials from either the particular materials' own properties or by comparison with other materials: toxicity can only be assessed on a case-by-case basis. Weaknesses of earlier studies reviewed in the report include poor characterization of the materials being studied, so that it is difficult or impossible to rely on the results to draw inferences about any other materials however similar they may superficially appear to be. No epidemiological studies had been carried out at all.

The FAO Expert Report identifies the situation as one of acute lack of information, made worse by the lack of any central databases co-ordinating and collating such data as is being generated. In particular, the potential for nanoparticles to re-enter the food chain through recycling either of food products or their packaging means that the life-cycle analysis for nanotechnologies in food is very complex. However, while the potential hazards remain scientifically unassessed, and indeed no consensus has been reached as to the appropriate methodology for assessing whether risks exist in any given scenario, there is no sound basis for policy making: to take as a basis the anxieties of a number of consumer and environmental advocacy groups could lead to foregoing the benefits of useful technologies which in fact have little risk, or whose potential for benefit outweighs that for harm.

As in other sectors to which nanotechnology may in due course be applied, the FAO Expert Report identified the question in food and agriculture as being how

19. Although there is an argument as to whether or not these micelles, comprising encapsulated or aggregated nanoparticles but themselves being in the region of 300 nm, should be included in definitions of nanotechnology for which (as discussed in Ch. 1) 100 nm is frequently used in defining the upper limit of size.

materials should be treated which are approved or subject to standards in their bulk form, but which may display different characteristics due to their nanoscale application. Materials such as bulk silica are permitted food additives[20] but may never have been tested in the form of nanosilica. Further, permitted additives such as titanium dioxide may in practice contain a fraction at the nanoscale even when manufactured on the bulk scale, but again there has been little or no measurement of the size of this fraction or testing as to any different properties it displays compared to the bulk materials. In agriculture, the use of nanoscale materials or delivery systems may offer considerable benefits in terms of reducing the total quantity of known pesticides which need to be deployed to protect a given crop, or of known supplements added to animal foodstuffs to maintain healthy livestock, if better uptake and more targeted application can be developed. However, the difficulty of tracing such products either into food products or into the environment may mean that the regulations applicable to these uses have to be adapted.

A central conclusion of the FAO Expert Report was that possible gaps in the food standards setting procedures of the Codex Alimentarius Commission need to be identified and addressed. Predictably, the expert group agreed that a clear, internationally harmonized set of definitions is needed relating to the use of nanotechnology in food packaging, as additives, in nutrients and in agrichemicals such as improved pesticide delivery to plants. As a corollary, a procedure is also needed for classifying nanostructures in agrifood products that supports risk governance for engineered nanomaterials. The report also pointed to the need to consider the whole life cycle of engineered nanomaterials in agrifood applications, and to be aware of the potential for biopersistence in the agricultural sector and consequent potential exposure during production and use, and through possible contamination of agrifood produce, soil and water. There were no recommendations for regulation at this stage, but the report emphasized the need for continuing dialogue with stakeholders and further research to improve the understanding of the behaviour of nanomaterials in the human or animal body and the environment.

The meeting concluded that current risk assessment approaches used by FAO/WHO and Codex are suitable for engineered nanomaterials used in food and agriculture but additional safety concerns may arise owing to the characteristic properties of nanomaterials which need to be addressed. For instance, very high surface areas means that the properties of surface molecules dominate and may have novel features including as to interaction with other molecules present in foods. These effects and interactions need to be characterized. There are also concerns as to the *in vivo* behaviour of nanomaterials which are developed as delivery systems for drugs and nutrients, and which might in theory also enhance the uptake from the gastrointestinal tract of other substances which the body could otherwise safely eliminate.

There is also a need for transparency and traceability of nano-enabled products or engineered nanomaterials in food and agriculture in order to built public trust in their use. Mechanisms to support this should be identified.

20. See INS 531.

The experts agreed that it was necessary to consider the potential impact of nanomaterials entering the food chain indirectly through their presence in the environment – and take this into account in designing life-cycle management for nanomaterials. For instance, incineration of products incorporating nanoparticles may simply release those particles into the atmosphere where, like other airborne pollutants, they may enter the lungs of both humans and animals and have the potential to cause malfunction or disease.

The review of the state of knowledge in respect of the risks nanotechnology may pose to human and animal health strongly underlined the high degree of uncertainty on almost all aspects for the time being. The FAO Report therefore highlights the importance of clear communication – as to what is known, how risks are being assessed and the approach to managing the introduction of these new products and materials in the face of the prevalent uncertainty and anxiety which it engenders. Stakeholders, including the public, need to be involved in the process of investigation of risks.

A follow-up meeting, NANOAGRI 2010, was held in Brazil in 2010. Once again, there was no consensus as to regulation at this stage; information gathering and risk assessment remain the key issues given the present state of knowledge. Recommended future steps included creation of an open-access database or portal on the application of nanotechnologies in food and agriculture, including information on market access requirements in different countries, and work together with other organizations to review and develop an internationally accepted tiered approach for risk assessment of nanoparticles.

Participants discussed whether, when approved food additives and ingredients are reformulated at the nanoscale to confer new functional properties, such products should be subjected to a new risk assessment. Improved bioavailability of food supplements such as minerals and vitamins manufactured at the nanoscale may lead to redefining such regulatory concepts as acceptable daily intakes (ADI) or recommended daily intakes (RDI) in order to prevent risks of overdosing. The definitions of purity criteria will require information on the size and form of a substance. With regard to regulations covering food contact materials, expressing regulatory migration levels in mass per mass or volume will not take into account the possibility of changing toxicity profiles with increased surface areas and smaller size. Conversely, regarding nanoscale food contaminants there may be a need to revalidate human health limits such as provisional tolerable weekly intakes or tolerable daily intakes due to possible increased toxicity of nano-sized particle contaminants. However, in order to enforce any such limits, routine analytical methods to detect nanoparticles in foods will be needed and these have not yet been developed.

5.5 WORLD HEATH ORGANIZATION/IFCS

The WHO is jointly responsible with the FAO for the Food Safety Programme, but also has initiated projects of its own relating to governance of nanotechnology, as a topic within its Intergovernmental Forum on Chemical Safety (IFCS). At the Sixth Session of the IFCS in Dakar in 2008 the activities of the various IOMC members were reviewed and a Statement on Manufactured Nanomaterials was agreed. However, recognizing that the work of IFCS substantially overlapped with the more recent UN initiative of the ICCM, the Dakar Statement was not taken forward by IFCS but instead presented at the second ICCM where it was substantially adopted as a Resolution, discussed at 5.3.2 above. IFCS has now been integrated into the ICCM.

The WHO held its Fifth European Regional Conference on Environment and Health at Parma in March 2010 and reached agreement on a Declaration[21] which includes a commitment to act on the key challenges on environment and health, one of which was:

> (e) concerns raised by persistent, endocrine-disrupting and bio-accumulating harmful chemicals and (nano)particles.

Participants consequently called for an increase in research into the potential adverse effects of nanoparticles to evaluate possible harmful exposures, but no other action.

---

21. EUR/55934/5.1 Rev. 2, at <www.euro.who.int/__data/assets/pdf_file/0011/78608/E93618.pdf>.

Chapter 6
Regulation in Europe

6.1 INTRODUCTION: EXISTING LEGISLATION

6.1.1 The Commission's Gap Analysis

In 2008, the European Commission produced a review document assessing the applicability of existing European legislation to nanotechnology[1] (the Gap Analysis). Since applications of nanomaterials arise in a number of different industry sectors, the regulation of each of those sectors may apply to nanomaterials – and different regulations may apply to the same nanomaterial when it is used in a different sector. The Gap Analysis identified the relevant areas of legislation as being grouped under chemicals, worker protection, products and environmental protection, simultaneously applicable. The common purpose is to ensure that materials are neither made, used nor disposed of within the EU which are harmful to humans, either directly or indirectly (through consumption by food animals for example, or by damaging ecosystems). There are accordingly similarities of approach at a high level – requirements to notify, to test and monitor – but the details in each case are materially different.

The Gap Analysis was intended to provide a road map for future legislative work insofar as it could relate to nanomaterials. The conclusion was that overall current legislation largely covered risks relating to nanomaterials but that it could need modification as new information becomes available. Specifically, the Commission recognized that the appropriate threshold values for triggering tolerance limits or safety reviews might be different where a substance was in nanoform from the limit applicable to the same substance in bulk form. But the Commission

1. *Regulatory Aspects of Nanomaterials* 17 Jun. 2008 COM(2008) 366 final, with annexed Staff Document.

concluded that there was as yet insufficient knowledge about nanomaterials to propose any changes. The major difficulty foreseen in the Gap Analysis was the implementation of any regulatory instrument (whether current or future) since the guidance and methodologies supporting risk assessment would have to be reviewed in order to ensure that they effectively address risks associated with nanomaterials and make best use of the information becoming available. Accordingly, the Commission pointed out that EU-funded research on nanomaterials which could help to fill some of the knowledge gaps was already in progress, and recommended that:

> Commission working groups, meetings of Competent Authorities and Agencies in charge of coordinating the implementation of regulation will have to examine on an ongoing basis whether and what type of further action is needed. These activities will mainly be reflected in documents that support implementation of existing legislation. Examples are the setting of thresholds, authorisation of substances and ingredients, qualifying waste as hazardous, reinforcing conformity assessment by reclassification, introducing restrictions on the marketing and use of chemical substances and preparations.

This conclusion, which could be characterized as 'cautious optimism' – nothing is fundamentally wrong and we are simply looking to improve the position further – did not satisfy the European Parliament.

6.1.2 The Political Response

Following further reports and opinions by its own committees, the European Parliament passed a Resolution[2] (the Parliament's 2009 Resolution) noting the inherent tensions in the subject:

> D. whereas nanomaterials . . . potentially present significant new risks due to their minute size, such as increased reactivity and mobility, possibly leading to increased toxicity in combination with unrestricted access to the human body, and possibly involving quite different mechanisms of interference with the physiology of human and environmental species,
> E. whereas the safe development of nanomaterials can make an important contribution to the competitiveness of the European Union's economy and to the achievement of the Lisbon strategy,
> F. whereas the current discussion about nanomaterials is characterised by a significant lack of knowledge and information, leading to disagreement starting at the level of definitions . . . .

2. European Parliament resolution of 24 Apr. 2009 on regulatory aspects of nanomaterials (2008/2208(INI)).

The Parliament noted that the current research funded by the EU was biased towards innovation rather than the sort of testing and toxicology methodologies which would be required to enforce any regulation effectively. It opined that:

> Q. nanomaterials should be covered by a multi-faceted, differentiated and adaptive body of law based on the precautionary principle, the principle of producer responsibility and the polluter-pays principle to ensure the safe production, use and disposal of nanomaterials before the technology is put on the market, while avoiding systematic recourse to general moratoria or undifferentiated treatment of different applications of nanomaterials.

The Parliament

> [did] not agree, before an appropriate evaluation of current Community legislation, and in the absence of any nano-specific provisions therein, with the Commission's conclusions that a) current legislation covers in principle the relevant risks relating to nanomaterials, and b) that the protection of health, safety and the environment needs mostly be enhanced by improving implementation of current legislation, when due to the lack of appropriate data and methods to assess the risks relating to nanomaterials it is effectively unable to address their risks.

The Parliament's 2009 Resolution therefore called on the Commission to review all relevant Community legislation within two years with a view to ensure safety for all applications of nanomaterials in products with potential health, environmental or safety impacts over their life cycle, and to ensure that legislative provisions and instruments of implementation reflect the particular features of nanomaterials to which workers, consumers and/or the environment may be exposed. Accordingly, a second regulatory review is scheduled for release at the end of 2011.

The Parliament also called on the Commission to compile before June 2011 an inventory of the different types and uses of nanomaterials on the European market and report on their safety, while respecting justified commercial secrets such as recipes, and to make this inventory publicly available. No direction was given as to how this was to be achieved given the effective dearth of such information in the public domain. On the knowledge gap issue, the Parliament's solution was to call for the proposal, as soon as possible, of the establishment of a permanent and independent European network responsible for monitoring nanotechnologies and nanomaterials, and a basic and applied research programme on the methodology for this monitoring (particularly metrology, detection, toxicity and epidemiology).

As this last solution illustrates, the Parliament does not itself have power to propose legislation; that is the prerogative of the Council and the Commission. Consequently this Resolution has no legal effect; indeed, it follows an earlier resolution[3] which emphasized the need for investigation of the effects of

---

3. Resolution of 28 Sep. 2006 on nanosciences and nanotechnologies P6_TA(2006)0392.

nanoparticles that are not readily soluble or biodegradable, in accordance with the precautionary principle, before such particles are put into production and placed on the market, which had no effect. The 2009 resolution has, however, increased the pressure on the Commission to accelerate its review of existing legislation, and has resulted in a number of revisions across the various sectors, some of which have been passed within the proposed two year deadline, others of which remain in draft or under discussion. The Parliament has also been active in proposing nanotechnology-specific clauses for any draft legislation which has come before it for review. The remainder of this chapter reviews the relevant fields of regulation and discusses the revisions which have been made or are or may be in preparation. Since each of these areas of regulatory law could in itself form the subject of a substantial book, the reviews are necessarily somewhat high level.

6.2 CHEMICALS (REACH AND CLP)

6.2.1 Overview

Chemicals in the EU are regulated under a single comprehensive set of regulations, Regulation 1907/2006 on the Registration, Evaluation, Authorisation and Restriction of Chemicals (REACH),[4] which came into force on 1 June 2007. It applies to all new substances, and since December 2010 is being applied progressively over a phase-in period to substances which were already on the market in the EU before it came into effect. All chemical substances will be regulated under REACH by the end of 2018.

REACH sets up, as its name implies, a system whereby chemical substances cannot be put on the market in the EU unless they have been registered and, where necessary, authorized – which may involve the imposition of restrictions. The system is very wide in scope, covering all substances whether manufactured, imported, used as intermediates or placed on the market, either on their own, in preparations or in articles. The process of registration, evaluation and authorization is centrally managed by the European Chemicals Agency (ECHA), aiming to ensure that the legislation is properly implemented and has credibility with all stakeholders. ECHA co-ordinates the substance evaluation process by the competent authorities of the Member States, and generally takes the resulting decisions,[5] which may include restrictions ranging from conditions on the manufacture, placing on the market or use of certain dangerous substances to an outright prohibition. This is intended to act as a safety net to manage Community wide risks that are otherwise not adequately controlled. ECHA also provides expert opinions to the Commission in the authorization and restriction procedures and is responsible for managing the confidentiality of and access to information provided by applicants.

4. OJ L 396, 30 Dec. 2006, 1.
5. Where the Member States' representatives disagree, there is a mechanism whereby the Commission would decide.

It also handles requests for exemptions from the registration requirement for product and process oriented research and development.

The system of registration and authorization requires the manufacturer or importer of a substance to provide to ECHA a dossier with specified information about the substance. Since this potentially involves many manufacturers carrying out identical tests on the same substance, ECHA assists in the formation of Substance Information Exchange Fora (SIEFs) for sharing data between applicants dealing with the same substance, particularly animal test data at the pre-registration stage. Industry consortia have also been established to reduce the amount of redundant testing carried out, which usually allow members to use available data and registration dossiers.

If a company fails to register a substance then the company is no longer allowed to manufacture or import this substance.[6]

6.2.2 Exemptions from REACH

The Regulation does exempt certain substances that are adequately regulated under other legislation,[7] like medicinal products, or that generally present such low risks as not to require registration, like water, oxygen, certain noble gases and cellulose pulp. Likewise, naturally occurring substances such as minerals, ores and ore concentrates, cement clinker, and so on, are not required to be registered as long as they are not chemically modified.

Polymers are exempted as well from the requirement to register for the time being, since they usually are not very hazardous, but in certain circumstances monomers in polymers have to be registered. This issue was the subject of one of the earliest legal challenges under the new regulations.[8] Article 6.3 of REACH requires registration of 'the monomer substances which compose a polymer' that have not already been registered. The ECJ took a cautious approach and ruled that this must be interpreted as applying only to reacted monomers, which are integrated into polymers. As a consequence, unreacted residual monomers in a polymer are regarded as impurities and are not part of the monomer registration process.

Other exceptions are substances which are radioactive, subject to customs supervision or are non-isolated intermediates.[9] Waste is specifically exempted. Food that meets the definition of a substance, on its own or in a preparation, will be subject to REACH; however, such substances are largely exempted from registration, evaluation and authorization under this legislation. They may

---

6. The 'no data, no market' principle at Art. 5 REACH.
7. Annexes IV and V of REACH.
8. *R (on the application of SPCM SA, CH Erbsloh KG, Lake Chemicals and Minerals Ltd, Hercules Inc.) v. Secretary of States for the Environment, Food and Rural Affairs* Case C-558/07, ECJ decision 7 Jul. 2009.
9. These categories are regulated under other instruments.

require authorization under food-specific instruments. Chemicals used in the manufacture of cosmetics are covered, again subject to exceptions where sector-specific legislation applies.

6.2.3 Registrability of Nanomaterials under REACH

There is no express reference in REACH to nanomaterials. Nevertheless, there is no doubt that they are covered by it. The key issue is the definition of 'substance', which is given, in very broad terms, in Article 3(1):

> Substance: means a chemical element and its compounds in the natural state or obtained by any manufacturing process, including any additive necessary to preserve its stability and any impurity deriving from the process used, but excluding any solvent which may be separated without affecting the stability of the substance or changing its composition.[10]

Materials which are registrable at all under REACH will, therefore, all be registrable if presented in nanoform. The remaining question is whether substances already registered in bulk form require additional and separate registration in nanoform. The short answer is, there is nothing in REACH stating that multiple registrations are required for any given chemical substance. However, the definition of substance is singular, but encompasses a chemical element and its compounds either in the natural state or manufactured. Since it appears arguable that a compound having different properties from the constituent elements would be registrable in addition to any registration of the individual elements, or that a compound produced by one process leaving a particular impurity profile could be registrable separately from the same compound produced by a different process resulting in a different impurity profile, then it is highly arguable that different nanoforms of a registered substance should likewise be subject to separate registration.

The ECHA's guidance indicates that once it is established that the substance falls within the normal scope of REACH, different provisions may apply according to the hazard properties associated with the particle size of the substances. The European Commission has also taken this view. In its Gap Analysis, it stated that when an existing chemical substance, already placed on the market as bulk substance, is introduced on the market in a nanoform, the registration dossier will have to be updated to include specific properties of the nanoform of that substance. The additional information will include different classification and labelling of the nanoform and additional risk management measures. The risk management measures and operational conditions will have to be communicated to the supply chain.

Thus, the ECHA may require additional registration for known materials when introduced in nanoform. This suggests that a nanoform of a material which in bulk form was on the market in the EU prior to REACH, and which therefore in bulk

10. This is identical to the definition used under the 7th Amendment of the Dangerous Substances Directive (Directive 92/32/EEC amending Directive 67/548/EEC).

form benefits from the transitional 'phase-in' arrangements, will not itself have the benefit of that timetable. This leads to questions about whether the properties of the nanomaterial should be included in the registration of the bulk substance if they are different, and whether the two sets of data are comparable if nano- and bulk substances are tested with different test methods. The inclusion of nano- and bulk form data may lead to contrasting properties and even conflicting classifications within the same dossier.

The need for such an approach was illustrated by a 2009 report commissioned by the Dutch Ministry of Housing, Spatial Planning and the Environment[11] (the Dutch Nanosilver Report) which described a hypothetical application for registration of nanosilver under REACH on the basis of publicly available information, throwing up a number of areas where the properties of nano- and bulk-forms of a chemically identical material – elemental silver – were not adequately described by considering a single set of data. For instance, particle size distribution is a different issue when considered within a bulk material context than a nanoparticle one, where particles outside the nanoscale are likely to be present but may be classed as impurities rather than simply 'more of the substance'. Dissolution characteristics are likely to differ at nano- and bulk scales. The issue of mass versus specific surface area as an appropriate metric for assessing the amount of material being put on the market, and for exposure scenarios for both worker and environmental exposure; the question of whether the state of agglomeration or aggregation of particles should be taken into account; and what parameters can be taken as an indicator of sameness of properties, potentially allowing read-across of data from the bulk form registration to any nanoform registration, need to be considered. The lack of databases on the effects of nanomaterials in the body means it is impossible for correlations to be drawn from *in vitro* test results to *in vivo* toxicity. There is a dearth of information on the kinetics of nanomaterials – their ability to move across boundaries within the body – which is not standard information where bulk materials are concerned but essential for the nanoscale. Different exposure routes – transdermal, as well as oral and inhalation – may be relevant. As a result, the Dutch Nanosilver Report proposed a risk assessment framework for first generation nanomaterials under REACH[12] which would remove the current tiering of testing requirements according to tonnage made or imported, instead imposing a base set of information, including detailed exposure assessment, required independently of any manufacture threshold and whether or not the same material was already on the market in a bulk form.

6.2.4 Registration Thresholds

Manufacturers and importers of substances are required to submit a registration to the ECHA for each substance manufactured or imported in quantities of 1 tonne or

11. *Nanomaterials under REACH: nanosilver as a case study* RIVM report 601780003/2009, MEJ Pronk et al.
12. Dutch Nanosilver Report, s. 6.

above per year, whether in isolation, as part of a preparation (mixture) or even incorporated into an article if, in the latter case, the substance is expected to be released under normal or reasonably foreseeable conditions of use.[13] Accordingly, unless less than 1 tonne is to be put on the market per year, no nanomaterials can be placed on the market in the EU without registration.[14] Further, if the substance is registered under the bulk form but the manufacturer proposes also to deal in the nanoform, it is subject to registration of the combined tonnage, as well as the deadlines applicable to the bulk substance.

This weight threshold has been the subject of considerable debate, given that many nanomaterials are used in very small quantities. Environmental and consumer groups are concerned that the threshold is inappropriate since the effects of nanoparticles may not be directly correlated with the mass produced, and such thresholds may effectively exempt many nanomaterials from REACH regulation entirely.

Although much debated, however, the threshold may not be so critical in practice. The REACH Competent Authorities Sub-group on Nanomaterials (CASG Nano) and Commission officials have taken the view that safety information will still be required on nanomaterials under the Classification, Labelling and Packaging Regulation (discussed at 6.2.10 below) regardless of tonnage. The information would then have to be included in Safety Data Sheets (SDS) for dissemination to all downstream users of the material.

It has been suggested that in its Review of REACH, due in 2012, the Commission may propose lowering the 1 tonne threshold so far as nanomaterials are concerned, possibly to the 10 kg threshold required under Directive 67/548/EEC (Dangerous Substance Directive), but this remains to be seen.[15] The thematic study on nanomaterials which forms part of the review process, 'Scientific technical support on assessment of nanomaterials in REACH registration dossiers and adequacy of available information' is being carried out by the JRC whose reports are due in September 2011 to November 2012.[16]

6.2.5 Substances of Very High Concern

In any event, the threshold does not apply to any nanomaterial whose safety data categorizes it as a Substance of Very High Concern (SVHC). These are substances which are: carcinogenic; mutagenic; toxic for reproduction; or persistent, bioaccumulative and toxic (as defined in Annex VIII of REACH); or which do not meet

13. Article 7 REACH.
14. Article 5 REACH.
15. Nanomaterials: Definitions and registrations under REACH, Smith, at <www.wca-environment.com/nanomaterials-definitions-and-registrations-under-reach/>, 24 Nov. 2010.
16. Notes from 8th Meeting of Competent Authorities for REACH and CLP, 8–10 Jun. 2011; at <http://export.gov/europeanunion/static/8CaracalFinal%20Report_Latest_eg_eu_036537.pdf>.

the definition but for which there is scientific evidence of probable serious effects to human health or the environment which give rise to an equivalent level of concern. SVHC substances have to be authorized and are likely to be restricted under REACH regardless of what volume is used. Additional requirements apply to this category – regardless of the total mass manufactured or imported – unless, in all cases, the producer can exclude exposure to humans or the environment during normal or reasonably foreseeable conditions of use including disposal, and supplies appropriate instructions to the recipient of the article.

ECHA publishes a list containing SVHC substances which have been put forward for registration, available online.[17] No nanomaterials are on the list at the time of writing, but should information become available which suggests that a nanomaterial may be hazardous within this category, then it may be added. If so, applicants will have to demonstrate that risks associated with uses of these substances are adequately controlled or that the socio-economic benefits of their use outweigh the risks. Applicants must also analyse whether there are safer suitable alternative substances or technologies and, if so, prepare substitution plans. The Commission may amend or withdraw any authorization on review if suitable substitutes become available.

6.2.6 Registration and Pre-registration

Since REACH applies to all substances including those already marketed in the EU, the process for registration is being staggered to assist both manufacturers/importers[18] and the ECHA to deal with the applications. The deadline for registration for substances of which 1,000 tonnes per year, or carcinogenic, mutagenic or toxic for reproduction (CMR) substances of which a tonne or more per year, are being sold has already passed on 1 December 2010. Substances of which 100–1,000 tonnes per year are being sold will have to be registered by 1 June 2013, while substances of which only 1–100 tonnes per year are being sold must be registered by 1 June 2018.

Manufacturers or importers of substances for which the deadline has not yet passed must pre-register certain basic information such as the name of the substance (including its European Inventory of Existing Commercial Chemical Substances (EINECS) and any Chemicals Abstract Service number); the registrant's name and contact person; envisaged deadline and tonnage for registration.[19] The information can be submitted online through the REACH-IT portal. For existing

17. <http://echa.europa.eu/chem_data/authorisation_process/candidate_list_table_en.asp>.
18. A non-EU manufacturer can under Art. 8 REACH appoint an EU legal person as 'Only Representative' to comply with the registration requirements on its behalf. The EU-based customer, which would otherwise be an importer under REACH and have to register, may thus not have to register, while the manufacturer retains confidentiality over its materials vis-à-vis the supply chain downstream.
19. Article 28 REACH.

manufacturers and importers, this information had to be submitted by 1 December 2008; for first time manufacturers and importers the information must be submitted within six months of their supply reaching the 1 tonne threshold, and not later than 12 months before the actual registration deadline. If the threshold is met later than this the company must proceed directly to register and must suspend manufacture/ import until three weeks after the registration dossier has been submitted. Pre-registration does not, however, constitute a legally binding commitment to register and continue to supply the substance.

Even after a substance has been registered, the manufacturer or importer is obliged to update the dossier as new information comes to light such as new risks to health or the environment, or as the volume of production or import changes.[20]

6.2.7 Identification of Substances under REACH

ECHA has produced Guidance on various aspects of the information required for registration, such as the identification and naming of substances.[21] The identification parameters for substances with unknown or variable compositions, complex reaction products and biological materials are very complex and detailed, and it is unhelpful that the Guidance fails to address issues of scale, stating baldly that 'a substance is completely identified by its chemical composition, the chemical identity and the content of each constituent in the substance'.[22] Later draft Guidance produced by CASG Nano on registration of nanomaterials in REACH instead stipulated that where either hazard properties of the nanoform differ from the bulk form, or physical properties of shape, structure or surface morphology of various nanoforms differ significantly from each other, then a separate classification and labelling as well as a separate dossier may be justified for the registration of a substance.[23]

Despite the wealth of guidance on testing bulk chemicals, the absence of validated methodologies for characterizing materials on the nanoscale means that manufacturers are having to undertake research into analytical techniques in order to comply. The list of known techniques is expanding steadily through the work of the OECD WPMN and the UN Committee of Experts on the Globally Harmonized System of Classification and Labelling of Chemicals. A list of possible measurement methods transmitted by the expert from Australia was considered at the twentieth UN Experts' session in December 2010,[24] and the results of Phase I of the OECD Sponsorship Programme on characterizing reference nanomaterials should become available in the course of 2011. Three projects aimed at

20. Article 22 REACH.
21. Guidance for identification and naming of substances under REACH. Available at <http://guidance.echa.europa.eu/docs/guidance_document/substance_id_en.pdf>.
22. Section 4.1.
23. CASG Nano/1/2008.
24. ST/SG/AC.10/C.4/2010/19.

adapting the existing technical guidance to the specific issues raised by nanomaterials were launched in 2009 and are scheduled to be completed in the course of 2011. RIPoN 1, on substance identification, has developed a number of case studies[25] intended to identify the relevant characteristics which determine the identity of a specific nanomaterial by assessing the adequacy of current substance identification parameters set down in Annex VI.2 of REACH. Unfortunately, the expert group held divergent opinions on a number of issues such that, rather than state an agreed approach, the report[26] describes a number of options and defines further necessary research. Further information is expected to be garnered from the 11 applications for registration which were received in the application period ending on 1 December 2010, which may be used to assist in evaluation of later registrations. However, gaps and uncertainties will remain for several years until a sufficient body of technical knowledge has built up. The ECHA comments[27] that the evolving science of nanotechnology may necessitate further requirements in the future to reflect the particular properties of nanoparticles.

6.2.8 The Dossier

As explained above, Article 10 REACH requires manufacturers and importers to submit:

- a technical dossier, including intrinsic properties, uses and exposures, for substances manufactured or imported in quantities of 1 tonne or more; and
- a chemical safety report, for substances in quantities of 10 tonnes or more.

Article 10 refers to voluminous identification and test data more particularly specified in Annexes VI-X REACH; preparation of the dossier is not a simple matter. Detailed guidance is available from the ECHA.[28] Much of this data may have to be obtained from companies further up the supply chain; thus, preparation for registration under REACH has affected much of the global chemicals industry. Testing can include physico-chemical, toxicological and ecotoxicological data; the amount of testing required depends on the tonnage concerned, within bands. Any data produced by testing which has to be carried out on vertebrates (which is only where other adequate data gathering methods are not available) must be shared with other applicants. Sharing of other test data is also strongly encouraged. Once the dossiers are submitted, the substance is registered – although this does not preclude the ECHA from raising objections later.

The threshold for provision of a chemical safety report was objected to by the European Parliament's 2009 Resolution, which called on the Commission to

25. Titanium dioxide, silver, CNTs, calcium carbonate.
26. Expected to be published in 2012.
27. <http://echa.europa.eu/help/help_faq_en.asp#FAQ>.
28. *Guidance on registration,* Version 1.6 (January 2011) at <http://guidance.echa.europa.eu/docs/guidance_document/registration_en.pdf?vers=31_01_11>.

introduce a requirement for a chemical safety report with exposure assessment for all registered nanomaterials.

Even if a substance is not subject to registration requirements, where the substance:

- meets the criteria for classification as dangerous in accordance with the Dangerous Substances Directive or Directive 1999/45/EC (the Dangerous Preparations Directive); or
- is persistent, bioaccumulative and toxic or very persistent and very bioaccumulative as measured by criteria set out in Annex XIII; or
- is included in the list of substances which are candidates for inclusion in Annex IV (substances whose use is prohibited unless specifically exempted or authorized);[29]

then the supplier must provide the recipient of the substance with a Safety Data Sheet in accordance with Annex II.[30]

The recipient may also provide information about the substance back up the supply chain to enable the supplier to prepare an exposure scenario for the use the recipient is making of the substance.

6.2.9 Register of Nanomaterials?

REACH does not provide for any register specific to nanomaterials, but in view of the lack of effectiveness of the various voluntary reporting schemes which have been tried[31] a number of countries have considered introducing a mandatory register. The European Parliament's 2009 Resolution included a call for the Commission to consider notification requirements for all nanomaterials placed on the market on their own, in preparations or in articles, which could be interpreted as the establishment of a Community-wide list of marketed nanomaterials. This will be under consideration in the REACH review due in 2012.

Meanwhile in June 2011 France's national Decree on the creation of a mandatory register for nanomaterials[32] being marketed at 100 g per year or more was submitted to the European Commission for scrutiny against EU internal market provisions. The threshold was increased to 100 g per year after consultation on a draft with a 10 g threshold, in order to exclude research activities from the register.[33] Registration is automatic, there is no intent to use the information to

29. Articles 57–59 REACH.
30. Article 31(1) REACH.
31. See Ch. 4.
32. *Decree on the annual declaration of nanoparticle substances, laid down in implementation of Article L. 523-4 of the Environmental Code.*
33. *French mandatory Reporting of Nanomaterials notified to the European Commission*, Nanotechnology Industries Association, at <www.nanotechia.org/global-news/french-mandatory-reporting-of-nanomaterials-notified-to-the-european-commission>.

restrict or prohibit marketing, it is merely intended to increase the information available.[34] If such a register falls within REACH's aim of harmonizing the requirements on manufacture, placing on the market or use of nanomaterials, then a separate national register may be a barrier to trade and incompatible with European law. If the Commission rules that the register cannot be established, then the issue may have to be referred to the Court of Justice of the EU for determination. This is likely to take several years. Pending this France, Italy and Belgium are undertaking a project comparing country approaches which aims to outline a proposal from the common experience as a basis for new national databases so that such projects are harmonized throughout the EU.

6.2.10 Classification and Labelling

Regulation 1272/2008/C on classification, labelling and packaging of substances and mixtures (the CLP Regulation) implements the second edition of the United Nations Globally Harmonised System of classification and labelling of chemicals (GHS) into EU law. The GHS is developed and maintained at United Nations level with the aim of avoiding different hazard information requirements on physical, health and environmental hazards for the same chemicals around the world. This is intended to facilitate trade: by applying GHS across different countries: it will no longer be necessary for the exporter of a chemical to reclassify and relabel it in order to comply with different classification criteria, rules and Safety Data Sheet requirements of the importing country.

The CLP Regulation came into force on 20 January 2009 and over a transitional period up to 2015 progressively replaces the Dangerous Substances Directive and the Dangerous Preparations Directive. ECHA is required to establish a European Classification and Labelling Inventory (C&L Inventory), a database that will contain basic classification and labelling information on notified and registered substances as well as the list of harmonized classifications.[35]

Manufacturers, importers, distributors and users of chemicals have to classify, label and package chemicals in accordance with the CLP requirements before placing them on the market in the EU. 'Placing on the market' includes the provision of free samples. The classification is to be based on the available information, including information created under REACH testing, and may also impose a requirement for additional testing of physico-chemical properties if necessary to determine whether or not a substance or mixture entails a health or environmental hazard as defined in Annex I.[36] A Safety Data Sheet is required for industrial customers if the product includes hazardous substances or mixtures which contain hazardous substances above a certain threshold. Guidance on the preparation of

34. Notification No. 2011/0307/F-COOC.
35. Table 3.1 of Annex VI to CLP Regulation.
36. Article 8(1) CLP Regulation.

SDS is in preparation at the time of writing.[37] From 1 June 2015 all SDS will have to use the CLP classification and labelling; until then, both Dangerous Substances Directive classification and, where appropriate, the CLP version must be used.

Notification to the C&L Inventory is required for:

- the manufacture or import of 1 tonne or more per year of substances (including isolated intermediates), or import of articles containing substances, which are subject to registration in accordance with REACH; or
- the manufacture or import of substances which are classified as hazardous or mixtures containing hazardous substances above the specified concentration limit, irrespective of the quantity involved,

unless the substance or mixture is in the finished state and intended for the final user or for uses for which there is specific legislation in place.[38] Notification is required within one month after the substance is placed on the market, even if the substance is one for which the REACH notification deadline does not arise until 2018. Unlike REACH, for hazardous substances it is *not* subject to an annual tonnage threshold,[39] although the substances to which it applies are the same categories as defined in REACH. Instead, the concentration limit is the threshold for notification. The concentration limit is either that laid down in Annex I of the CLP Regulation or, where scientific information shows either that the substance is hazardous below that level or that it is *not* hazardous at a level above the Annex I limit, that set by the manufacturer, importer or downstream user.[40] Where the information which would form the notification has already been supplied to the ECHA as part of a registration, a duplicate notification is not required.[41]

Substances need to be classified only in the form that is going to be placed on the market and in which they can reasonably be expected to be used. Factors to consider include particle size, agglomeration or aggregation state or crystal structure – and accordingly a change to any of these which may affect the classification will trigger a further evaluation of the revised form.[42] Since not all nanoparticle formulations represent a more pronounced hazard than the bulk form of the same substance, the characterization has to be case by case pending a more complete knowledge of the properties of nanomaterials. If the particle size may have a significant effect on the test result, the test report for the relevant hazard should explicitly state what the particle size is. If several particle sizes are placed on the market or where the particle size may be altered during transport or storage, a worst-case approach should be used. This would normally imply using the

37. the current draft is available at <http://guidance.echa.europa.eu/docs/draft_documents/fourth_draft_sds_Guidance_july_2011.pdf>.
38. Such as radioactive materials, medicinal products, cosmetic products and food and feed.
39. Article 39(b) CLP Regulation.
40. Article 10 CLP Regulation.
41. *How to Notify Substances in the Classification and Labelling Inventory,* at <http://echa.europa.eu/doc/publications/practical_guides/pg_7_clp_notif_en.pdf>.
42. Classification, labelling and packaging of nanomaterials in REACH and CLP, CA/90/2009 Rev 2, 3 Dec. 2009.

classification based on testing the smallest particle size that could occur. Further, if particle size is relevant for classification and safe handling and use, this should be mentioned in the Safety Data Sheet, as should information on deviating classifications due to different particle sizes.

A Safety Data Sheet must be provided and must be updated when either new information on the hazards posed by the substance become available or if a substance not formerly having been classified as dangerous is so classified. The CLP Regulation also requires manufacturers and importers to classify substances that are subject to registration or to notification in line with Article 7 or 9 of REACH, even if not placing them on the market. This includes for instance the classification of substances that are used for product and process-orientated research and development.

6.2.11 Confidential Information in the REACH System

REACH itself contains no provisions for the protection of confidential information, but the risk that this could be compromised through either the data sharing or registration requirements is clearly recognized.

Article 118 relates to 'Access to Information' held by ECHA, and begins by confirming that the principles on transparency in Regulation 1049/2001 apply. Article 4 of that Regulation require an EU institution to withhold a document in response to an access request if disclosure would undermine any person's commercial interests. Article 118(2) of REACH specifically acknowledges that the disclosure of details of the full composition of a preparation; precise use, function or application of a substance or preparation; precise tonnage of substances and preparations; or links between a manufacturer or importer and downstream user may undermine the protection of the commercial interests of the concerned persons. Thus, in principle they are protected from disclosure. Nevertheless, ECHA is entitled to disclose such information if urgent action is essential to protect human health, safety or the environment, such as in emergency situations.

A registrant can request that components of the information in the dossier, which would ordinarily be made available online, be treated as confidential.[43] The request must include a justification that is accepted by ECHA as to why publication of this information is potentially harmful to commercial interests (of either the requestor or of any other involved party).

ECHA also recommend that companies participating in a Substance Information Exchange Forum (SIEF) or other data sharing arrangement should protect the confidentiality of their information either by entering into confidentiality agreements limiting access to documents or other information to specific named persons, or departments, for example, only the persons working within a regulatory section are allowed to see certain information. This access could, for instance, be confined only to access to a physical copy without the ability to copy it. If the

43. Article 10(a)(xi) and Art. 119.2 REACH.

information is too confidential to share with competitors even on such a basis, then the documents could be reviewed by a third party consultant who would report to the other SIEF participants but not show them the information.

Further, a registrant can 'opt-out' from the joint submission of data if doing so would lead to disclosure of confidential information.[44] However, a party which opts out is still bound by the data sharing obligations.

6.3 PHARMACEUTICALS, PESTICIDES AND BIOCIDES

6.3.1 PHARMACEUTICALS

6.3.1.1 Overview

Few areas of industry are so tightly regulated as pharmaceutical production and sales. From a tradition in which medicines were expensive, poorly understood and often as likely to kill as cure, the developed countries have swung to the opposite extreme. Whereas the quack doctors of the nineteenth century were left to practise upon the credulity of their patients, pharmaceutical companies today are expected to comply with standards of ethics which require the disclosure of any information which may throw light upon the possible effects of their products. Medicinal products are subject to prior governmental approvals as to the conditions under which they are produced, the uses to which they can be put, the information which must be supplied alongside them, and the way the products are labelled and packaged. Pharmaceutical companies are also required to track the after-effects of their products and report any unexpected adverse reactions.

No medicinal product can be marketed anywhere in the EU without a marketing authorization (MA).[45] 'Medicinal product' includes any substance or combination of substances presented for treating or preventing disease in human beings, or which may be administered to human beings with a view to making a medical diagnosis or to restoring, correcting or modifying physiological functions in human beings. In addition to artificially prepared compounds it includes micro-organisms (e.g., vaccines) and naturally occurring substances such as plant extracts. It applies equally to products which are said to be medicinal 'by presentation' as to those which are medicinal 'by function'. All such products must be authorized by the competent government or EU agency.

In order to obtain an MA, the applicant must demonstrate the product's safety, efficacy and quality to the required standards. Demonstrating safety and efficacy involves very extensive pre-clinical and clinical testing, all of which must itself be

44. Article 11.3(b) and 19.2(b) REACH.
45. Article 6 Directive 2001/83/EC on the Community code relating to medicinal products for human use and Art. 3 Regulation (EC) No. 726/2004 laying down Community procedures for the authorization and supervision of medicinal products for human and veterinary use and establishing a EMA.

conducted in accordance with stipulated standards and procedures. The data required includes relevant clinical (human pharmaco-kinetics, safety and efficacy data, any post-marketing experience) and non-clinical (pharmacology, toxicology, pharmaco-kinetics) studies. To demonstrate that the product is of the necessary quality, data must be supplied showing the manufacturing processes to be used for the product, their stability, characterization and control.

6.3.1.2 Nanomedicine

Accordingly, whereas other industry sectors are having to introduce new practices in order to establish that any nanomaterial included in their products is safe, the pharmaceutical and medical device industry is already set up to conduct such studies. Similarly, the regulatory authorities handling applications for MAs have well-established technical committees to assess the data provided. Many products now included under the label 'nanotechnology' have been in use in medicine for many years, such as nanoscale liposomes for drug delivery and large soluble nanoscale molecules such as recombinant peptides or oligonucleotides, which have been manufactured using well-established techniques. As discussed in Chapter 1, it is not appropriate to consider these within the category of nanotechnology. Indeed, the October 2011 Recommendation on a definition of nanomaterials specifically acknowledges that:

> (17) Given the special circumstances prevailing in the pharmaceutical sector and the specialised nano-structured systems already in use, the definition in this Recommendation should not prejudice the use of the term 'nano' when defining certain pharmaceuticals and medical devices.

This is not to say that there may not be issues which still need to be considered. The regulatory approach to nanopharmaceuticals (principally drug delivery systems such as liposomes and nanoparticles) will largely be determined by the active moiety/ payload rather than the delivery system. A number of such systems have already been approved under the existing regulatory structure, where the standard processes used for their manufacture are well-described and understood. Examples include the formation of mixed micelles (liposomal doxorubicin) or colloidal dispersions. However, the delivery system may have its own effect on the pharmaco-kinetics of the drug, for instance through aggregation/agglomeration, surface composition, charge or reactivity. This may be positive – there have been reports that nanoscale particles have an affinity for solid tumours resulting from their enhanced permeability and retention effect. But it could also be negative, for instance through giving rise to immunological effects, as a result of surface modifications, which the bulk forms do not show. As with industrial chemicals, the question arises of whether nanopharmaceuticals require a new authorization when the 'bulk' forms of the same chemical substance are already authorized. Further, it is possible that nanoforms may gain access to tissues and cells which larger particles may not, and may therefore have novel effects. Their clearance from the body may also take different pathways.

Medical devices are also being developed on the nanoscale, such as nanostructure scaffolds for tissue replacement, nanostructures that allow transport across biological barriers, remote control of nanoprobes, integrated implantable sensory nanoelectronic systems and multifunctional chemical structures for drug delivery and targeting of disease.

6.3.1.3 Regulatory Response

The European Medicines Agency's (EMA) Committee on Medicinal Products for Human Use (CHMP) produced a reflection paper in 2006, acknowledging that:

> Although nanosizing does not necessarily imply novelty, it is expected that nanotechnology will yield innovative products. Such products could span the regulatory boundaries between medicinal products and medical devices, challenging current criteria for classification and evaluation. Appropriate expertise will need to be mobilised for the evaluation of the quality, safety, efficacy and risk management of nanomedicinal products and the need for new or updated guidelines will be reviewed in the light of accumulated experience.[46]

The EMA has created an Innovation Task Force to ensure EMA-wide coordination of scientific and regulatory competence in the field of emerging therapies and technologies, including nanotechnologies, and to provide a forum for early dialogue with applicants on regulatory, scientific or other issues that may arise from the development of such products. Such requests for scientific advice are an established procedure, for instance in designing proposed clinical trials and defining the trial protocol. The procedure can be valuable if the applicant asks specific scientific questions, giving its current position on each of them and justification for it, with reference to the supporting data. It is most valuable in product areas where there is no guidance, or if national advice shows a lack of consensus. But the applicant will be expected to take account of the answers given, so it is imperative to seek the advice at a stage where amendments to the trial are still feasible, and certainly not once the trial is in progress.

In September 2006, the UK's Medicines and Healthcare Products Regulatory Agency (MHRA) Commission on Human Medicines conducted a survey of the literature relating to studies on nanoparticles in healthcare.[47] At that time, no chronic data was identified. The Commission concluded, however, that bulk chemical toxicity need not be predictive of toxicity at the nanoscale and the relevance of past nanotoxicology reviews is limited for the nanotechnology used in healthcare products. Further, even at the nanoscale safety data obtained on particles with a diameter of 100–1000 nm, which includes most existing nanoscale drug delivery vehicles, is not always directly transferable to chemically identical

46. *Reflection Paper on Nanotechnology-Based Medicinal Products for Human Use*, EMEA/CHMP/79769/2006.
47. *The toxicology of nanoparticles used in healthcare products* at <www.mhra.gov.uk/home/groups/dts-bi/documents/websiteresources/con2025205.pdf>.

particles at diameters of 1–100 nm. Accordingly, the mechanism behind the size dependency needs to be known to decide what accuracy in sample characterization is needed.

The Commission did observe that dermal penetration of nanoparticles appears to be restricted to anionic nanoparticles and also dependant on the charge concentration; particle size is therefore only indirectly a factor. There were at the time no non-biodegradable healthcare products on the market, so it was not surprising that not enough toxicity data was found to allow any conclusions. However, within the data that was available, the Commission did not identify any mechanism of toxicity in healthcare nanoparticles that would evade conventional hazard identification testing currently required to comply with the regulations for healthcare products.

6.3.1.4 Application for a Marketing Authorization

There are three routes for obtaining an MA: the centralized procedure, administered by the EMA for the whole of the European Economic Area (EEA); the decentralized procedure, which must be used for any product not subject to the centralized procedure and not yet authorized anywhere in the EEA for which approval is sought in more than one Member State; and the mutual recognition procedure, consisting in principle of recognition by the regulators of all EU Member States of an MA issued by one of them. The decentralized procedure is fundamentally similar to the mutual recognition procedure in principles though not in timing.

The centralized procedure will apply to nanotechnology medicinal products only if the applicant shows that the medicinal product constitutes a significant therapeutic, scientific or technical innovation or the granting of authorization under the centralized approach is in the interests of patients or animal health at Community level.

Whichever route is followed, the fundamental requirements are the same: the application must demonstrate the quality, safety and efficacy of the product through reference to sufficient, credible data. Consequently, the MA comprises not only any document issued to confirm that an authorization is granted and specify any particular conditions, but also the entire data dossier submitted with the application for authorization, and associated documents.

The data included in the dossier are protected from disclosure to third parties, and in particular may not form the basis for an application from a third party for authorization of the same product until a specified period, now harmonized at 10 years,[48] has elapsed since submission.

Once an MA has been granted and the product is placed on the market, it will need to be reviewed after five years for the risk/benefit balance to be reassessed. If

48. Subject to extension in certain special cases.

after that assessment the MA is renewed it will continue in effect indefinitely. An MA for a product which is not marketed within three years of being approved will lapse, unless there are exceptional circumstances leading the competent authority to waive the lapse. For instance, litigation over an allegation of patent infringement might be a legitimate reason for the MA holder to elect to delay marketing the product for such a period of time.

Throughout the life of a medicinal product the MA holder is required to take into account technical and scientific progress, which would include new knowledge becoming available as to any difference in effect between a nanoparticle of an active and the bulk form of the same, and to make any amendments that may be required to the MA. They also have the option of altering or improving the product on their own initiative or adding additional safeguards to its use.

6.3.1.5 Post-marketing Pharmacovigilance

Ongoing pharmacovigilance is essential even once an MA has been granted, since the data upon which an approval is based are necessarily generated from a very small patient population over a relatively short time period. Different effects may be observed when the same product is used with the population as a whole, both as regards interactions with other medications, different patient characteristics and simply the statistical observability of extremely rare reactions. Hence the approval process includes the need to submit a pharmacovigilance plan. The MA holder should continue to consult all known sources of additional information about the product, including literature searches, liaison with the regulatory authorities and the AEGIS database, as well as having all medical representatives alert for possible reports of adverse effects from both patients and doctors.

In addition, periodic safety update reports are required to allow periodic assessment of a product's risk-benefit profile. In the EU, these are done every six months for the first two years, annually for the next three years and every three years thereafter.

Post-authorization, the regulators retain the right to inspect in order to confirm continued compliance with quality requirements and pharmacovigilance procedures. Both the EMA and national regulators have these rights; in practice until September 2011 national regulators have been more thorough in the exercise of them, carrying out detailed reviews of plant, processes and interviews with relevant personnel.

6.3.1.6 Pharmaceuticals and CLP

Finished medicinal and veterinary products and medical devices are not covered by the CLP Regulation. But where a manufacturer or importer supplies substances and mixtures, for example, active pharmaceutical ingredients (APIs) or excipients, that are not yet in the finished state, this manufacturer or importer will have to classify, package and label these substances and mixtures in accordance with CLP.

In addition, if these substances are placed on the market, they will also have to be notified to the C&L Inventory.

This exemption from the provisions of the CLP Regulation does not distinguish between active and non-active pharmaceutical ingredients: it applies to any substance or mixture used in medicinal products, including excipients, which is in the finished state and intended for pharmaceutical use.

6.3.2 PESTICIDES

6.3.2.1 Principles

The EU system for regulation of pesticides is based upon a two-tier registration system. Regulation (EC) 1107/2009 concerning the placing of plant protection products on the market[49] (the Pesticides Regulation) requires active ingredients to be assessed at Community level for inclusion on a positive list (known as Annex 1). Products containing chemicals listed on Annex 1 must then be assessed, as to issues that were not considered in the assessment for inclusion in Annex 1, and registered by Member States. No product can be placed on the market in a Member State unless it has been approved and registered in accordance with this system.[50]

In order to be included in Annex 1 a substance must be shown to be without unacceptable risk to people or the environment. All existing active substances which were already on the market before this system was introduced have now been reviewed for inclusion in (or exclusion from) Annex 1.

The Pesticides Regulation is expressly based upon the precautionary principle. Article 1(4) states:

> The provisions of this Regulation are underpinned by the precautionary principle in order to ensure that active substances or products placed on the market do not adversely affect human or animal health or the environment. In particular, Member States shall not be prevented from applying the precautionary principle where there is scientific uncertainty as to the risks with regard to human or animal health or the environment posed by the plant protection products to be authorized in their territory.

This would appear to give Member States considerable discretion as to whether to permit or prohibit the marketing in their territory of pesticide products even where the active substances have been included in Annex 1. However, given the high standard of safety which must be demonstrated for such inclusion, it would be very difficult for a Member State to prohibit a product unless there was reason to believe that the formulation or delivery system posed a threat which was not present in the active taken in isolation.

---

49. OJ L 309/1, 24 Nov. 2009.
50. Article 28 Pesticides Regulation.

6.3.2.2 Procedure

In order for an active substance to be added to Annex I, the manufacturer must apply with a complete dossier of data. The data requirements have been effectively harmonized at OECD level and the *OECD guidelines and criteria for industry for the preparation and presentation of complete dossiers and summary dossiers for plant protection products and their active substances in support of regulatory decisions in OECD countries* must be followed. The complete dossier is a very extensive set of documentation comprising not only the required tests and studies, but also a series of supporting documents providing background information on the active substance and its uses. A set of 'tiered' summaries detailing the applicant's evaluation and risk assessment for the active substance and product is also required, along with full copies of all the individual study reports.

To enable a risk assessment to be conducted, at least one plant protection product must be supported by the dossier with comprehensive data to support the proposed use or uses.

The dossier is evaluated by a single Member State, which produces a structured draft assessment report (DAR) comprising an evaluation of all the submitted studies together with a risk assessment and overall proposal for inclusion or non-inclusion of the substance in Annex I. The DAR is submitted to the European Food Standards Agency (EFSA) and undergoes detailed scientific scrutiny of the evaluation and risk assessments by experts drawn from Member States and EFSA, as required. EFSA may also refer issues for further consideration and resolution to its independent Panel on Plant Protection Products and their Residues (PPR). EFSA presents its conclusions on the risk assessment to the European Commission, which will produce a proposal for inclusion or non-inclusion depending on the conclusions of the risk assessment process and considering the risk management options available. This is subject to a vote by Member States followed by adoption and publication.

6.3.2.3 Nanotechnology and Pesticides

There is no reference to nanotechnology in the Pesticides Regulation, but EFSA has developed a practical approach for assessing potential risks arising from applications of nanoscience and nanotechnologies in the food and feed chain.[51] This guidance covers the physico-chemical characterization requirements of engineered nanomaterials used as pesticides and testing approaches to identify and characterize hazards arising from the nanoproperties. It does, however, acknowledge the incomplete state of scientific knowledge and is therefore liable to be updated as understanding increases.

51. Guidance on the risk assessment of the application of nanoscience and nanotechnologies in the food and feed chain EFSA Journal 2011; 9(5):2140.

6.3.3 Biocides

6.3.3.1 Existing Provisions

The current non-agricultural pesticides (biocides) regime is regulated by Directive 98/8/EC.[52] Like the pesticides regime, it deals with the authorization and placing on the market of biocidal products in the EU, and the establishment at Community level of a list of active substances which may be used in biocidal products. There are detailed provisions as to the 'proper use' of biocidal products. Proper use includes measures necessary to keep the use of biocidal products to a minimum as well as an obligation to ensure that their use in the workplace complies with the directives on health and safety protection for workers.[53] The authorization system is based on the principle of mutual recognition of authorizations, in which disagreement between Member States as to whether or not a product should be authorized is resolved by a Standing Committee on Biocidal Products.

Marketing authorization may be granted for a product only if its active substances are listed in the Annexes to the Directive[54] and the requirements set out in the Annexes are met. A dossier must be supplied with data establishing that the biocidal product is sufficiently effective, has no unacceptable effects on the target organisms, human or animal health or on surface water, groundwater or the environment. Biocidal products classified as toxic, carcinogenic and mutagenic or toxic for reproduction will not be authorized for marketing to the general public.

Even once listed, biocides must be classified, packaged and labelled in accordance with the Dangerous Substances Directive and Dangerous Preparations Directive, which applies until 1 June 2015. Authorizations may be reviewed at any time during the period for which they have been granted, and the holder of an authorization is responsible for passing on immediately to the competent authority any information of which the holder is aware concerning an active substance or a biocidal product containing it and which may affect continuing authorization.

The applicant may ask for certain sensitive information to be kept confidential from all persons other than the competent authorities and the Commission. Some information, however, such as the name and address of the applicant, the physical and chemical properties of the biocidal product, cannot be included in this confidentiality clause.

Member States may use the information contained in the dossier accompanying the application for authorization for the benefit of another applicant only on certain conditions, one of which stipulates that the written agreement of the first applicant must be obtained. If such agreement has been obtained, then the subsequent applicant can rely on the earlier data provided that they can adduce evidence that the product is similar and its active substances are the same as in the

52. Directive of the European Parliament and of the Council of 16 Feb. 1998 concerning the placing of biocidal products on the market (as amended) OJ L 123, 24 Apr. 1998, 1.
53. See s. 6.6 below.
54. The maximum period during which a substance may be listed in the Annex is 10 years.

product previously authorized. The applicant and holder or holders of former authorizations are encouraged to come to an agreement on the shared use of information to avoid duplicating the testing on vertebrate animals.

Specific information must be included in a data safety sheet provided by the person placing the product on the market, to enable professional and industrial users of biocidal products to take the necessary measures for the protection of the environment and health.

6.3.3.2 The Proposed New Regime

This regime is the subject of a pending proposal for revision, the second reading for which in the European Parliament was scheduled for October 2011, with a vote in early 2012. The draft revised Directive[55] includes references to nanomaterials in biocides for the first time. Recital 64 of the current draft states:

> There is scientific uncertainty about the safety of nanomaterials for human health and the environment. In order to ensure a high level of consumer protection, free movement of goods and legal certainty for manufacturers, it is necessary to develop a uniform definition for nanomaterials, if possible based on the work of appropriate international fora, and to specify that the approval of an active substance does not include the nanomaterial form unless explicitly mentioned. The Commission should regularly review the provisions on nanomaterials in the light of scientific progress.

As this wording shows, there remains a reluctance to commit even to a definition of nanomaterials subject to amendment, the solution which had been adopted in the Draft Novel Foods Regulation.[56] The European Parliament had at its September 2010 reading defined 'nanomaterial' to mean

> any intentionally produced material that has one or more dimensions of the order of 100 nm or less or is composed of discrete functional parts, either internally or at the surface, many of which have one or more dimensions of the order of 100 nm or less, including structures, agglomerates or aggregates, which may have a size above the order of 100 nm but retain properties that are characteristic of the nanoscale. Properties that are characteristic of the nanoscale include:
>
> (i) those related to the large specific surface area of the materials considered; and/or
> (ii) specific physico-chemical properties that are different from those of the non-nanoform of the same material.

55. Text as at 21 Jun. 2011 available at <http://register.consilium.europa.eu/pdf/en/11/st05/st05032-re02.en11.pdf>.
56. Discussed at s. 6.4.7 below.

The European Council version going forward to the second reading refers instead to the newly adopted[57] Commission Recommendation concerning the definition of nanomaterials. The Recommendation defines nanomaterial as:

> a natural, incidental or manufactured material containing particles, in an unbound state or as an aggregate or as an agglomerate and where, for 50% or more of the particles in the number size distribution, one or more external dimensions is in the size range 1 nm-100 nm.
>
> In specific cases and where warranted by concerns for the environment, health, safety or competitiveness the number size distribution threshold of 50% may be replaced by a threshold between 1 and 50%.

However, recital 16 to the Recommendation specifically precludes the definition from prejudging or reflecting the scope of application of any piece of Union legislation or of any provisions potentially establishing additional requirements for those materials, including those relating to risk management. It acknowledges that it may in some cases be necessary to exclude certain materials from the scope of application of particular legislation even if they fall within the definition, or to include additional materials, such as some materials with a size smaller than 1 nm or greater than 100 nm in the scope of application of specific legislation suited for a nanomaterial. There thus remains some uncertainty as to what form may eventually be used in the revised Biocides Directive.

In Article 4 (Conditions for approval) of the current draft, the inclusion of nanoforms by default where the same chemical substance has been approved in bulk form is expressly ruled out:

> The approval of an active substance shall not cover nanomaterials except where explicitly mentioned.

This is again less specific than the European Parliament's proposal, which had been that a biocidal product including nanomaterials could be authorized only if the risk to the environment and to health had been assessed separately. Nor does the Council position include the Parliament's proposal for labelling of any ingredient in nanoform with the word 'nano'.

The Polish presidency hopes to reach an informal Political Agreement on the text by the end of 2011, and a formal second reading deal at the Environment Council in January 2012. If agreed, the revised directive is likely to come into effect in September 2013. However, as with all regulatory provisions around nanomaterials this one is also subject to ongoing revision as the state of knowledge improves. The current draft acknowledges the potential gaps in testing methodologies so far as nanomaterials are concerned, stating:[58]

57. Discussed at s. 1.2.6 above.
58. At Annex VI on Common Principles for the Evaluation of Dossiers.

In the case of biocidal products containing nanomaterials the principles set out in this Annex will also need to be adapted and elaborated in technical guidance to take account of the latest scientific information.

6.4 FOOD AND FEEDS

6.4.1 OVERVIEW

The European Food Safety Authority (EFSA), established in 2002 under Regulation 178/2002[59] (the Food Regulation), has responsibility for maintaining a high level of protection of human life and health and consumers' interests in relation to food, while also ensuring free movement of food and feed within the EU through uniformity of food requirements between the various Member States. It provides independent scientific advice through a Scientific Committee consisting of a series of panels dealing with particular topics, and support for Community legislation and policy affecting food and feed safety.

The safety of most foodstuffs depends entirely on the chemical composition of the product, which (where products of animal origin are concerned) may in turn depend on the feed consumed by the animal. Food and feed regulation is therefore concerned with the entire food chain, from the production and harvesting of the raw materials through their processing, packaging and distribution to the consumer. So far as nanotechnology is concerned, at the time of writing this is regulated solely under the general provisions relating to food production since, although in principle nanomaterials could be introduced by way of malicious contamination of food (as has sometimes happened by persons introducing non-food materials such broken glass into food processing lines), this is rare and a question for the criminal law rather than any matter of food regulation. Specific regulation of nanomaterials in food is under discussion in the form of a new draft novel foods regulation (see section 6.4.7 below) but has not yet been finalized.

The issue of nanotechnology in food and animal feeds, however, is a particularly contentious one since, whereas hazard assessment for inhalation of fine particles is relatively well understood, the interaction between fine particles and the human or animal gut is very much less explored. Very little is known about the ability of the gastrointestinal tract to absorb or reject nanomaterials, nor has their absorption, distribution, metabolism and excretion (ADME) from the body been studied. Finally, there are no validated methods for the detection of nanomaterials in food and feed.

Accordingly, the emergence of new potential food substances or additives incorporating nanoparticles or nanoscale delivery mechanisms poses a challenge for EFSA. Environmental and consumer organizations' response to these

59. Regulation laying down the general principles and requirements of food law, establishing the European Food Safety Authority and laying down procedures in matters of food safety OJ L 31, 1 Feb. 2002.

uncertainties is to advocate amending the legislative provisions and the various EFSA Guidelines for the safety evaluation of substances for foods, food contact materials and food additives to include a definition of nanoscale, development of nano-specific testing methods prior to any nanoform substances being authorized and identification of nanoscale ingredients in the list of ingredients on packaged foods.[60] In response, guidance was issued in 2011 and is discussed in detail at 6.4.3 below.

6.4.2 Principles of Food Safety

Modern European food policy derives from the European Commission's Green Paper on the General Principles of Food Safety and the Communication of 30 April 1997 on Consumer Health and Food Safety.[61] Where a possibility of harmful effects on health has been identified but a full risk assessment, taking into account all relevant risk factors, including technological aspects, the best available scientific evidence and the availability of inspection sampling and testing methods, is not possible, the Green Paper proposed the application of the precautionary principle, requiring policy makers to balance the potential for harm to animal and human health and the environment from any new product or process against the perceived benefits. This precautionary approach is however only a fall back position: a proper risk assessment is strongly preferred.

The Food Regulation applies to all stages of production, processing and distribution of food other than private domestic food production and preparation.[62] Food includes drink, chewing gum and any substance or product, processed or otherwise, which is either intended to be or reasonably expected to be ingested by humans. Accordingly, a nanomaterial which is intended or expected to be ingested through having been incorporated into a food product, for whatever purpose, is covered by the definition. Feed,[63] live animals, unharvested plants, medicinal products, tobacco and narcotics are excluded.[64]

The fundamental provision is Article 14, which prohibits the placing on the market of unsafe food, meaning food either injurious to health or unfit for human consumption. This very broad provision applies whether the food is made in the EU or imported into it,[65] and also applies to food manufactured purely for export.[66]

60. Discussed in *Review of Nanomaterial and Nanoproduct Regulation* Working Group 3 of the Nanokommission, September 2010, at <www.bmu.de/files/english/pdf/application/pdf/nano_abschlussbericht3_en_bf.pdf>.
61. COM(97) 176.
62. Food Regulation Art. 1(3).
63. By Art. 3, a feed is any substance or product, including additives, whether processed or unprocessed, which is to be used for oral feeding to animals.
64. Food Regulation, Art. 2.
65. Food Regulation Art. 11.
66. Food injurious to health by EU standards cannot be exported even with the express agreement of the country of import.

Similarly, by Article 15 feed cannot be marketed or fed to a food-producing animal if it is unsafe, defined as having an adverse effect on human or animal health or making the food produced from the fed animal unsafe for human consumption.

The concept of 'unsafe' as regards food under Article 14 is tied to the information provided to the consumer, whether on the label or otherwise, concerning the avoidance of specific adverse health effects. A food is thus not unsafe even if the consumer suffers harm, if the consumer has been properly informed how to avoid the risk – for instance, has been informed that raw meat needs to be cooked before it can safely be eaten – but has disregarded the information.[67] Consequently, labelling becomes paramount where potentially harmful materials are concerned.

The concept of 'injurious to health' is to be considered by reference not only to immediate, short- or long-term effects on the consumer but also on subsequent generations. Thus, a nanomaterial which might be capable of affecting the reproductive organs or gametes without harming the consumer would still be classed as injurious to health. Once a hazard has been identified, the food operator must assess the risk regardless of whether or not the substance concerned is specifically regulated[68] – the absence of a directive on nanomaterials does not mean that they are exempted from risk assessment. Cumulative toxic effects must be taken into account, and also the particular health sensitivities of any specific category of consumers for whom the food is intended – such as those with particular food intolerances or allergies.[69]

Article 17 places responsibility for ensuring that foods and feed meet these standards with the producers and marketers of those products,[70] but Member States are also required to monitor and verify that the standards are being fulfilled at all stages of production, processing and distribution. The Article 17 obligations are directly applicable, but the implementation of the system is through national laws and supervisory arrangements, so that although an overall joint responsibility is imposed on the food chain, individual operators' liability is still governed by a combination of contractual arrangements and national legislation. The implementation is intended to apply hazard analysis critical control point principles.

These simple principles are easy to abide by where the risk to health is established and understood, but many circumstances exist in which this may not be so. One of the classic failures of the European regulatory system has been the emergence of new diseases transmissible from humans to animals such as Mad Cow Disease (BSE/vCJD) and avian flu, or potential risks from the use of genetically modified organisms. Potential risks from nanomaterials or technologies incorporated into food could fall into the same knowledge gap. Accordingly, Article 7 of the Food Regulation permits temporary measures to be taken based on the precautionary principle:

---

67. Food Regulation Art. 14(3).
68. Guidance of the Standing Committee on the food chain and animal health, 26 Jan. 2010.
69. Food Regulation Art. 14(4).
70. Potentially including charities as well as commercial operators, although where food is produced or marketed by a charity on an intermittent basis – occasional fund-raising events, perhaps – then this requirement may not apply.

1. In specific circumstances where, following an assessment of available information, the possibility of harmful effects on health is identified but scientific uncertainty persists, provisional risk management measures necessary to ensure the high level of health protection chosen in the Community may be adopted, pending further scientific information for a more comprehensive risk assessment.
2. Measures adopted on the basis of paragraph 1 shall be proportionate and no more restrictive of trade than is required to achieve the high level of health protection chosen in the Community, regard being had to technical and economic feasibility and other factors regarded as legitimate in the matter under consideration. The measures shall be reviewed within a reasonable period of time, depending on the nature of the risk to life or health identified and the type of scientific information needed to clarify the scientific uncertainty and to conduct a more comprehensive risk assessment.

This would permit EFSA to promulgate rules governing the use of nanomaterials in food and feed if health effects had been identified, pending a comprehensive risk assessment. No such measures have yet been introduced. EFSA's position is that a risk assessment is required not only in the context of the authorization of regulated substances but also where nanoparticles may be present as a contaminant in food or feed.

EFSA published a scientific opinion on nanoscience and nanotechnologies in relation to food and feed safety in 2009,[71] in which it concluded that established approaches to risk assessment can be applied to engineered nanomaterials but that a case-by-case approach was necessary. It acknowledged that in practice the current limitations of available data and lack of validated test methodologies would make risk assessment of specific nano-products very difficult and the outcomes uncertain. As a consequence, the Scientific Committee recommended specific areas for study, such as the interaction and stability of engineered nanomaterials in food and feed, the gastrointestinal tract and biological tissues; methods for detection and characterization of engineered nanomaterials in food, feed and food contact materials; and the development and improvement of test methodologies for the assessment of toxicity. It was not surprising, in view of these conclusions and recommendations, that in the first opinion adopted on a particular nanomaterial, silver nanoparticles in silver hydrosol, the Panel on Food Additives and Nutrient Sources concluded that it did not have enough information to be able to assess whether they are safe or not.[72] However, the Panel on the Safety of Food Contact Materials has since adopted an opinion concluding that a specific use of titanium nitride nanoparticles in plastic drinks bottles does not give rise to toxicological

---

71. Scientific Opinion of the Scientific Committee on a request from the European Commission on the Potential Risks arising from Nanoscience and Naontehcnologies on Food and Feed Safety *The EFSA Journal* (2009) 958, 1–39.
72. Scientific Statement of the Panel on Food Additives and Nutrient Sources added to Food on a request from the Commission on silver hydrosol as a source of silver added for nutritional purposes to food supplements, *The EFSA Journal* (2008) 884, 1–3.

concerns.[73] This was because the data available showed that titanium nitride as such is chemically inert and completely insoluble in all food simulants tested. Further, the Panel carried out diffusion modelling to investigate the potential migration of TiN nanoparticles into the food substance, and concluded from the results that no migration occurs.

6.4.3 EFSA Guidance on Engineered Nanomaterials

In May 2011 the Scientific Committee finalized its detailed *Guidance for the Assessment Concerning Potential Risks Arising from Applications of Nanoscience and Nanotechnologies to Food and Feed* (the Nanoscience Guidance),[74] relating to the presence of engineered nanomaterials. For this purpose, a broad definition of engineered nanomaterials is used: it includes any nanomaterial produced either intentionally or unintentionally (due to the production process) to be used in the food and feed area, with at least one size measurement between approximately 1 and 100 nm. The reference to unintentionally produced nanomaterials in particular means that many companies dealing in products which may have a particle size distribution as a result of the production process including a significant fraction in the range 1–100 nm will need to consider the impact of this Guidance on their operations.

The Nanoscience Guidance is based upon the EFSA scientific opinion of 2009, and sets out detailed requirements as to the information which will be needed in order to conduct a reliable risk assessment, proposing that applicants should provide studies able to:

- characterize the nanomaterial at the physico-chemical level, including any changes to the parameters in different environments: as manufactured (pristine state); in formulations delivered for use in food/feed products; as present in the food/feed matrix; as used in toxicity testing; and as present in biological fluids and tissues; and
- demonstrate its absorption, distribution, metabolism and elimination in the body, including whether it transforms into a non-nanoform in the food/feed matrix or in gastrointestinal fluids.

Six cases are presented which outline different toxicity testing approaches. Indicators that in depth testing will be necessary include:

- absorption and distribution leading to internal exposure;
- a high level of engineered nanomaterial reactivity or mobility;
- complex morphology;

---

73. Scientific Opinion of the Panel on food contact materials, enzymes, flavourings and processing aids on 21st list of substances for food contact materials *The EFSA Journal* (2008) 888–890, 1–14.
74. *The EFSA Journal* 9, no. 5 (2011): 2140.

- interactions with biomolecules such as enzymes, receptors, 'Trojan horse' effect;
- complex transformations (e.g., changes of surface properties, porosity) or metabolites;
- intended use as antimicrobials (e.g., unintended consequences on the gut flora).

In contrast, a loss of nano-specific properties following ingestion will move the risk assessment into a conventional risk assessment and the nano-specific risk assessment procedure will no longer apply. Nanostructured modifications on surfaces, and nanostructures that do not release particles and are not reactive are generally considered to reduce the likelihood of adverse effects (e.g., nanopores which can be used in filters).

Even where a nanomaterial transforms completely into a non-nanoform, however, information will still be required as to the timing and location of that dissolution/degradation, since until that moment the nanoform nature of the engineered nanomaterial may influence the biological behaviour, including kinetics and local effects.

The Nanoscience Guidance categorizes nanomaterials into those for which the relevant substance is already approved in bulk form for the same intended use and those for which it is not already approved.

For those which are already approved in bulk form for the same intended use, only supplementary and specific information is required on the potential additional hazards and risks that may arise from the nanoform. This will include comparison of information on ADME, toxicity and genotoxicity of the non-nanoform with, in first instance, ADME, repeated-dose 90-day oral toxicity study in rodents and genotoxicity information of the engineered nanomaterial. If the differences observed indicate increased hazard, then more toxicity testing will be required.

For those which are not approved in bulk form, where the engineered nanomaterial persists in the food/feed matrix and in gastrointestinal fluids, toxicity tests on the engineered nanomaterial should follow the relevant EFSA guidance for its intended use with some modifications in the testing due to the nanoproperties, as described in the Nanoscience Guidance. The modifications relate to the kinetics of the nanomaterial, and different measurement and testing protocols from the standard bulk form protocols may, therefore, be required.

Anticipated exposure scenarios from the proposed uses should be outlined. These will assist in evaluating the extent of the hazard characterization required, and provide parameters for the exposure assessment required for the risk assessment.

Appropriate *in vitro* and *in vivo* studies on the engineered nanomaterial should be undertaken, first to identify hazards and then to obtain dose-response data to characterize the hazards. But this uncontroversial principle still poses problems in practice, since the Nanoscience Guidance acknowledges that some test models and standard testing protocols used for non-nanoform substances may not necessarily be appropriate or optimal for the testing of engineered nanomaterial. The state of scientific knowledge is still the fundamental constraint.

This Guidance is obviously extremely valuable to companies wishing to use nanomaterials in the context of food, food contact materials or additives. However, the incomplete state of scientific knowledge means that there will still be uncertainties related to the identification, characterization and detection of engineered nanomaterial which are related to the lack of suitable and validated test methods to cover all possible applications, aspects and properties of engineered nanomaterial and other matters. The Guidance therefore recognizes that it will need to be updated based on experience and acquired knowledge – it will necessarily remain a 'work in progress' for some time to come.

6.4.4 Traceability

In addition to the requirement for prior safety assessment and authorization, any substance intended or expected to be incorporated into a food or feed must be established to be traceable at all stages of production, processing and distribution,[75] and all parties in the supply chain must be identifiable to the competent authorities on demand. In practice, this means that food operators must be able to identify one step above them in the chain for all their inputs, and one step below, maintaining records of suppliers, products and customers[76] as well as dates and times of transactions or deliveries and the volume or quantity of materials supplied. Operators are also encouraged to keep batch number records. It is not strictly necessary to maintain internal traceability, recording which ingredients went into which products, but EFSA considers it helpful if operators can do so, and this is likely to be particularly encouraged in the case of nanomaterials since their identification once incorporated is more difficult than that of bulk substances.

The effectiveness of this system in avoiding harm to consumers has been demonstrated by the fact that an accidental dioxin contamination of cattle foodstuffs, discovered through routine testing on a Dutch farm in 2004, was able to be traced and halted despite the feed having been distributed in Belgium, France and Germany as well as the Netherlands, before contaminated milk was sold for consumption.

6.4.5 Packaging

6.4.5.1 General Provisions

The traceability provisions do not cover materials used in packaging foodstuffs, but this is the subject of a separate regulation, Regulation 1935/2004 (the Food Packaging Regulation).[77] Like the Food Regulation, this regulation is in broad terms

75. Food Regulation Art. 18.
76. Other than at retail level.
77. OJ L 338/4, 13 Nov. 2004.

covering any material or article which is, is intended to be or can reasonably be expected to be brought into contact with food. Nanomaterials currently used for packaging or food processing or any other food contact purpose, such as packaging that acts as a barrier, or has coatings to block out moisture, oxygen or Ultraviolet (UV) light, are thus regulated.

Food contact materials are required to be safe. The basic principles are that components used in any food contact material should not transfer into the food in unacceptable quantities, a process known as migration. There are limits as to the overall migration of any substance which can migrate, which is 60 mg/kg food, and for individually authorized substances specific migration limits are set on the basis of the toxicological evaluation of the particular substance. No individual limits have been set to date for any material incorporating nanotechnology and so the overall migration limit applies for the time being.

This limit is a good illustration of the issue which has raised consumer concerns, since although 60 mg of any bulk material is a very small amount, if the material is in the form of nanoparticles it represents an enormous number of particles which may penetrate into tissues and have a biological impact many times greater than the same mass of the bulk substance could do. Despite this there do not appear to have been any proposals to date to revise this limit in respect of nanoparticles. However, revision of the EU legislation on plastics intended for food contact materials is currently in preparation and a Commission Regulation on this is anticipated soon. It is expected to address the regulation of nanomaterials specifically, clarifying that the nanoform of a substance is not covered by an authorization applied for and granted in respect of the bulk form of the same substance.

6.4.5.2 Active and Intelligent Packaging

Active and intelligent packaging is an area in which nanotechnology may bring particular benefits. Active packaging is defined as packaging intended to extend a food's shelf-life or maintain or improve the condition of packaged food by deliberately releasing into the food or absorbing from it. Intelligent packaging monitors the condition of the packaged food or the surrounding environment. The substances responsible for the active and/or intelligent function are still packaging even if contained in a separate container, for example, inclusion in a small paper sachet. Substances directly incorporated into the packaging material, for example, incorporation in the plastic of a plastic bottle are also covered and must be evaluated. In addition to the requirements of the Food Packaging Regulation, active and intelligent packaging components are also covered by Regulation 450/2009, and in particular may not be used to mask spoilage of food or mislead the consumer. In some cases, it may be necessary to evaluate and authorize the combination of substances, when the active or intelligent function implies interaction between different substances leading to an enhancement of the function or the generation of new substances responsible for the active and intelligent function.

Regulation 450/2009 introduced an authorization scheme for all such materials, which cannot henceforward[78] be used in food packaging unless they appear on a Community list[79] or within a short list of exceptions such as materials which only indirectly contact the food across a functional barrier.[80] However, there is a further exception to this exception in sub-section 5(c), paragraph (ii) of which relates to:

> substances deliberately engineered to a particle size which exhibit functional physical and chemical properties that significantly differ from those at a larger scale.

Accordingly, any packaging material incorporating engineered nanoparticles showing different properties from the bulk form will need to be included on the Community list. The manufacturer or importer of the substance(s) responsible for the active and/or intelligent function of the material will need to apply to EFSA demonstrating that under normal or foreseeable conditions of use it does not transfer its constituents to food in quantities which could endanger human health, bring about an unacceptable change in the composition of the food or bring about a deterioration in the organoleptic characteristics of the food,[81] unless within the limits imposed in the Directive on food additives.[82] EFSA will review the application and issue an opinion on each substance. In some cases, restrictions may be proposed by EFSA on a group of substances, especially when the active or intelligent function implies interactions between different substances.

6.4.5.3 Labelling and Advertising

In addition, intelligent materials and the presentation, labelling and advertising of food are further required not to mislead consumers.[83] All ingredients of a food must be indicated on the label[84] to enable consumers suffering from food allergies to avoid the relevant allergens. A '25% rule', which previously exempted from inclusion on the label any ingredient comprising less than 25% of the finished product, which would have included all nanomaterials, was abolished with effect from 25 November 2005.

6.4.6 Additives, Supplements and Fortification

6.4.6.1 Additives

Under Regulation (EC) No. 1333/2008 (the Food Additives Regulation), food additives must not be placed on the market in the EU unless they have been

78. All existing products had to be approved by 21 Jan. 2011.
79. To be maintained by EFSA.
80. Article 5 Regulation 450/2009.
81. Regulation 1935/2004 Art. 3.
82. Directive 89/107/EEC.
83. Food Regulation Art. 16 and Food Packaging Regulation Art. 4(4).
84. Directive 2000/13/EC as amended by Directive 2001/101/EC.

authorized for a given technological purpose following a comprehensive safety assessment.[85]

In view of the potential use of nanoparticles in the food area, priority was given to the issue of nanotechnology when the EU provisions on food additives were revised. The Food Additives Regulation therefore makes provision for re-evaluation of safety and, where appropriate, re-authorization of food additives used in a form that differs from the form previously assessed, which would include re-assessment where a known additive was introduced for the first time in nanoscale form.

If, for instance, additives are to be used in the form of micelles or similar, these must meet the requirements of the authorization procedure under Regulation 1331/2008/EC.[86] What particular tests are to be undertaken is left to be assessed on a case-by-case basis in the context of the application, which may of course lead to inconsistencies in the testing required and uncertainties as to the comparability of results, until a body of experience has been developed. Although clearly undesirable, it is difficult to see how the necessary knowledge could otherwise be built up unless through publicly funded research.

There are no labelling requirements under the Food Additive Regulation, but it is possible for the competent authority to impose conditions of use which could in appropriate cases include the obligation to label foods containing nanoscale additives with that information. As at 2009, the only nanoscale additives in use in the EU were flow-aids.[87]

6.4.6.2 Supplements and Fortification

Food supplements are governed by Directive 2002/46[88] which prohibits the marketing of any food supplement unless it consists of the vitamins and minerals listed in an Annex to the Directive in the form there listed. This is not, however, a reference to physical form but to which salt or form of chemical complex. The Directive therefore does not appear to affect the use of the approved forms of supplement in a nanoparticulate form or delivered through a nanoscale delivery system.

Similarly, Regulation 1925/2006 on the addition of vitamins and minerals and of certain other substances to foods[89] creates a positive list of vitamin and mineral nutrients that may be added to foods and specifies the forms in which those nutrients may be supplied according to their bioavailability and safety. As in respect of

---

85. Regulation (EC) No. 1333/2008 of the European Parliament and of the Council of 16 Dec. 2008 on food additives, OJ L 354, 31 Dec. 2008, 16–33.
86. Regulation of 16 Dec. 2008 establishing a common authorization procedure for food additives, food enzymes and food flavourings, OJ L 354, 31 Dec. 2008, 1.
87. EFSA Scientific Opinion, *The Potential Risks Arising from Nanoscience and Nanotechnologies on Food and Feed Safety*, 10 Feb. 2009.
88. Directive 2002/46/EC on the approximation of the laws of the Member States relating to food supplements OJ L 183/51, 12 Jul. 2002.
89. OJ L 404/26, 20 Dec. 2006.

supplements, the stipulation of the form of the substance used to fortify food appears to contemplate only the chemical rather than the physical form.

6.4.7 Novel Foods

6.4.7.1 The Existing Regime

As a matter of language, the existence of legislation relating to novel foods would appear directly relevant to the use of nanomaterials and nanotechnologies in food since these must be novel. Regulation 258/97 concerning novel foods and novel foods ingredients[90] (the Novel Foods Regulation) set up a system for prior authorization of any novel foods within its scope before they could be released onto the market, through a collaboration between Member States and the Commission. An additional Commission Regulation, 1852/2001, provided for the protection of information about a novel food which was submitted for evaluation during the authorization process. However, this first Novel Foods Regulation was passed in the light of public concerns over genetically modified organisms, well before nanotechnology was seriously contemplated in the food industry, and it is to these that it was primarily directed. Nevertheless, the definition of novel foods in Article 1(2) includes:

(c) foods and food ingredients with a new or intentionally modified primary molecular structure;

and

(f) foods and food ingredients to which has been applied a production process not currently used, where that process gives rise to significant changes in the composition or structure of the foods or food ingredients which affect their nutritional value, metabolism or level of undesirable substances.

Although there is nothing in these criteria specifically directed to particle size or properties arising from the nanoscale, the assessment procedure includes not only details of composition but also nutritional value, metabolism, intended use and level of chemical contaminants. If necessary in EFSA's opinion, additional studies of toxicology or allergenicity can be required.

The Regulation in its current form does therefore include the potential for detailed assessment of nanomaterials in foodstuffs. For instance, foods and other food ingredients used in foods in the form of micelles, which facilitate dispersal of fat-soluble components in water-based foods and vice versa, could fall within the scope of this Regulation. Even though the technology is not new, each case may need to be examined to assess whether the use of this technology gives rise to a significant change in the food or food ingredient and whether it was used prior to

90. OJ L 43, 14 Feb. 1997, 1–6.

15 May 1997. The lack of clarity of this definition was the focus of substantial comment from stakeholders in the Commission's consultation over a proposed revision of the Regulation; it was felt to raise questions over food production processes as well as the actual objective, of identifying changed foods.

6.4.7.2 Proposals for Revision

In 2002, discussions as to revision of this Regulation began, in particular as regards streamlining the authorization procedure and (separately) the treatment of traditional foods from third countries which under the existing version of the Regulation were treated as novel foods within the EU. In 2003, genetically modified organisms were removed from the scope of the Regulation by the passage of a separate Directive specifically focused on them.[91] In 2007, a draft replacement[92] (the Draft Novel Foods Regulation) was published with a view to improving the access of new and innovative foods to the EU market, while maintaining a high level of consumer protection and ensuring food safety. Under the Draft Novel Foods Regulation, a novel food would be subject to centralized authorization procedure and by Article 6 cannot be placed on the market unless:

(a) it does not, on the basis of the scientific evidence available, pose a safety concern to the health of the consumer under normal consumption conditions;
(b) it does not mislead the consumer, by the way it is presented or by its intended use;
(c) in the case where it is intended to replace another food, it does not differ from that food to such an extent that its normal consumption would be nutritionally disadvantageous for the consumer.

Thus, if information were available suggesting any health or safety concern from the use of nanoparticles in food, such particles could not be marketed. The European Parliament did not feel that this went far enough in respect of nanomaterials, however – it does not require any active testing to ascertain whether any concern should exist – and adopted at the second reading in 2010 a position which required that foods produced with the aid of nanotechnology may not be included in the Community list until specific methods have been approved for use, and an adequate safety assessment on the basis of these methods has shown that the use of the respective foods is safe.

The Commission will receive the application for authorization and EFSA will carry out the scientific assessment on the product. The 'one door – one key' approach will be taken for the approval of novel foods, as well as for food additives and food flavourings. This means that the applicant may make one application for approval covering all these possible uses of the substance in question.

91. Directive 19829/2003/EC.
92. COM(2007) 872 final 2008/0002 (COD).

Recital 6 of the Draft Novel Foods Regulation explicitly specified that foods modified by new production processes such as nanotechnology and nanoscience were intended to be within the scope of the provision. The objective was a systematic evaluation of food containing engineered nanomaterials, irrespective of whether nanomaterials were thought to cause changes in the food properties.

In the Draft, novel foods are defined[93] to include:

> (iii) food to which is applied a new production process, not used before 15 May 1997, where that production process gives rise to significant changes in the composition or structure of the food which affect its nutritional value, metabolism or level of undesirable substances.

This does not appear materially different from the existing definition in the existing Novel Foods Regulation, despite the criticisms. Nevertheless, the authorization system proposed under the Draft clearly includes any nanomaterial which is adjudged to have any of the listed effects. Accordingly, all such materials will become legally subject to the new centralized authorization procedure once the revised proposal is finalized and in effect.

Various definitions of nanomaterials were discussed for inclusion in the Draft. A generic version originated by the Commission's DG Sanco proposed that 'engineered nanomaterial' mean

> any intentionally produced material that has one or more dimensions of the order of 100 nm or less or that is composed of discrete functional parts, either internally or at the surface, many of which have one or more dimensions of the order of 100 nm or less, including structures, agglomerates or aggregates, which may have a size above the order of 100 nm but retain properties that are characteristic of the nanoscale.
>
> Properties that are characteristic of the nanoscale include:
>
> (i) those related to the large specific surface area of the materials considered; and/or
> (ii) specific physico-chemical properties that are different from those of the non-nanoform of the same material.[94]

This is the same as the definition proposed in September 2010 for use in the revised Biocides Directive. The approach did not reach consensus however. A further draft was proposed in October 2010, being:

> (1) 'Engineered nanomaterial' means any nanomaterial which is intentionally produced.
> (2) 'Nanomaterial' means a material that meets at least one of the following criteria:

93. Article 3(2)(a)(iii).
94. Council Common Position Art. 3(2)(c)).

(a) is made of particles, including aggregates or agglomerates thereof, with one or more external dimensions in the size range 1 nm–100 nm for more than 1% of their number size distribution; or
(b) has internal or surface structures in the size range 1 nm–100 nm; or
(c) has a specific surface area by volume higher than 60 m^2/cm^3.

(3) 'Particle' means a minute piece of matter with defined physical boundaries.

This is based upon the draft intended for use by the Commission in future legislation across all industrial sectors, as it stood at that time. The definition has changed substantially following consultation, and so it is likely that it will be materially different in any future draft of the Novel Foods Regulation. The Draft Novel Foods Regulation expressly contemplates that the definition should be open to adaptation as international standards and consensus emerge. In its present form, it would exclude materials which have an unavoidable trace number of particles at the nanoscale but which are not as to 99% of their constituent particles within that range. The reference to internal or surface structures in the range of 1–100 nm is intended to retain those nanomaterials in agglomerate or aggregate form whose overall dimensions may be several hundred nanometres but still possess properties associated with the nanoscale.

Authorization will require the person who is putting the material on the market in the Community to submit a request to the Member State in which the food is to be marketed, copied to the Commission. The request must include copies of the studies which have been carried out and any other material which may demonstrate the compliance of the food with the stipulated criteria: that it does not present a danger for or mislead the consumer, and that it does not differ from foods or food ingredients which it is intended to replace to such an extent that their normal consumption would be nutritionally disadvantageous to the consumer.[95] This information will then be assessed, within three months from receipt of a complete request[96] meeting the conditions; guidance is to be provided through EFSA to assist in particular small- and medium-sized enterprises in submitting applications.[97] The safety assessment is to be done under the same procedure as that for food additives, enzymes and flavourings (which are to be excluded from the Regulation, being covered by other Community legislation),[98] and the standard to be applied will take account of comparable food categories or the food which the novel food is intended to replace.[99] This may require the development of new testing procedures where existing techniques do not give unambiguous or sufficient information.

95. Article 3(1) of the Draft Novel Foods Regulation.
96. Which must also include an appropriate proposal for the presentation and of the product, and a summary of the dossier – Art. 6(1) of the Draft Novel Foods Regulation.
97. Article 9 of the Draft Novel Foods Regulation.
98. Discussed at s. 6.4.6 above.
99. Article 10 of the Draft Novel Foods Regulation.

Post-authorization market monitoring may also be required, and independently of such an obligation food producers have an ongoing obligation to inform the Commission if new scientific or technical information becomes available which might influence the safety evaluation, or if a third country imposes any prohibition or restriction on the novel food on its market.[100]

The data which is submitted in support of the application will remain exclusive to the applicant for five years,[101] so that second or subsequent applicants for authorization cannot rely upon it in support of their applications unless they obtain the agreement of the initial registrant.

The draft proposal was ultimately rejected by the European Parliament in March 2011, due to concerns over the inclusion of foods deriving from cloned animals. It will therefore not become law in its existing form and the existing Novel Foods Regulation remains in effect for the time being. The Commission has stated, however, that it intends to introduce a revised version as rapidly as possible to enable the provisions, other than those which provoked the previous controversy, to be passed and implemented in the near future. The revised text is expected to be released in late 2011.

6.4.8 Labelling

The Commission has also introduced labelling requirements for foods containing engineered nanomaterials as part of its ongoing legislative revision of food legislation. Regulation 1169/2011 on the provision of food information to consumers was adopted in October 2011.[102] The new regime requires food labels to specify any ingredients included at nanoscale, at the level of the list of ingredients. The existing food labelling Directive[103] had not achieved the objective of enabling consumers to make decisions readily concerning the foods they buy as the existing label information is generally found confusing. The mandatory requirements remain broadly the same under the new regime, with the option for the Commission to propose new requirements for specific issues. The revised legislation also covers nutrition labelling for foodstuffs, previously covered under a separate Directive.[104]

The final text includes at recital 25 the observation that:

> In order to inform consumers of the presence of engineered nanomaterials in food, it is appropriate to provide for a definition of engineered nanomaterials. Taking into account the possibility of food containing or consisting of

100. Article 11 of the Draft Novel Foods Regulation.
101. Article 12 of the Draft Novel Foods Regulation.
102. OJ L 304, 22 Nov. 2011, 18.
103. Directive 2000/13/EC of 20 Mar. 2000 on the approximation of the laws of the Member States relating to the labelling, presentation and advertising of foodstuffs.
104. Council Directive 90/496/EEC of 24 Sep. 1990.

> engineered nanomaterials being a novel food, the appropriate legislative framework for that definition should be considered in the context of the upcoming review of Regulation (EC) No 258/97 of the European Parliament and of the Council of 27 January 1997 concerning novel foods and novel foods ingredients.

Article 2(2)(t) defines an engineered nanomaterial as:

> Any intentionally produced material that has one or more dimensions of the order of 100 nm or less or that is composed of discrete functional parts, either internally or at the surface, many of which have one or more dimensions of the order of 100 nm or less, including structures, agglomerates or aggregates, which may have a size above the order of 100 nm but retain properties that are characteristic of the nanoscale.

Properties that are characteristic of the nanoscale include:

- (i) those related to the large specific surface area of the materials considered and/or
- (ii) specific physico-chemical properties that are different from those of the non-nanoform of the same material.

Finally, Article 18(3) requires all ingredients present in the form of engineered nanomaterials to be clearly indicated in the list of ingredients, by including the word 'nano' in brackets after the name of the ingredient.

The new Regulation is now in force. Compliance with these labelling requirements will be compulsory from 13 December 2014, save that by Article 54 stocks labelled before that date can continue to be marketed until the stock is exhausted.

6.5 COSMETICS

6.5.1 Overview

The history of cosmetics goes back several millennia, and unfortunately includes a number of instances where, in the avid pursuit of beauty, consumers have suffered significant harm through the use of toxic substances to whiten skin (lead in face powders) or dilate pupils (belladonna, or deadly nightshade) and so on. Thus, although there may be less anticipation of harm from the use of cosmetics than from pharmaceuticals, which could be expected to have stronger effects on the body, there is an evident need for appropriate regulation.

Cosmetic products are composed of a combination of chemical substances, whether derived from plant or animal extracts or synthesized from petro-chemical derivatives and other mineral sources. From the regulatory perspective, the latter are perhaps simpler to deal with having been synthesized from known constituents and thus being capable of precise chemical definition. Plant and animal extracts (shea butter; essential oils; and so on) however, are complex mixtures of complex organic molecules which may not be homogeneous, capable of precise analysis or consistent from one batch or harvest to the next.

6.5.2 The Current Regime

Currently, cosmetics are regulated in Europe under Directive 76/768/EEC[105] (the Old Cosmetics Directive). A cosmetic may be any substance or preparation intended to be placed in contact with the various external parts of the human body or with the teeth and the mucous membranes of the oral cavity, with a view exclusively or mainly to cleaning them, perfuming them, changing their appearance, and/or correcting body odours, and/or protecting them or keeping them in good condition.[106] In striking contrast to the position for pharmaceuticals and foods, where only substances which have been identified and approved can be marketed, any substance can be used as an ingredient in a cosmetic product as long as it is not prohibited.

The list of prohibited substances is set out at Annex II of the Old Cosmetics Directive, and at Annex III is a list of substances which cosmetic products may contain only under certain restrictions and conditions. Neither list includes any reference to particle size; accordingly, substances which are safe in their bulk form are not restricted in nanoform by the Old Cosmetics Directive.

There are limits to this liberal approach: Article 2 of the Old Cosmetics Directive provides that only cosmetic products which do not cause damage to human health when they are applied under normal or foreseeable conditions of use, can be put onto the market in the EU.[107] The manufacturer (formulator) is required to keep a dossier of information on its products readily accessible for the competent authorities of the Member States,[108] including the physico-chemical and microbiological specifications of both the raw materials and the finished product and an assessment of the safety of the finished product as regards human health. The Scientific Committee for Consumer Products (SCCP) has issued *Notes for Guidance* which suggest this could cover issues such as oral toxicity, mucous membrane irritation, skin absorption, reproductive toxicity, inhalation toxicity and phototoxicity and photomutagenicity.[109] Member States must

105. Directive of 27 Jul. 1976 on the approximation of the laws of the Member States relating to cosmetic products (OJ L 262, 27 Sep. 1976, 169).
106. Article 1, Old Cosmetics Directive.
107. Article 2, Old Cosmetics Directive.
108. Article 7a, Old Cosmetics Directive.
109. *The SCCP's Notes of Guidance for the testing of cosmetic ingredients and their safety evaluation*, 6th revision, 19 Dec. 2006, at <http://ec.europa.eu/health/ph_risk/committees/04_sccp/docs/sccp_o_03j.pdf>. The Introduction points out that these notes are concerned with testing and safety evaluation of the cosmetic ingredients listed in Annexes III, IV, VI, and VII of Dir. 76/768/EEC and those for which safety concerns have been expressed, but are also of interest to all cosmetic ingredients intended to be incorporated in a finished cosmetic product and can be of practical use in making a TIF (Technical Information File) or PIR

check the safety of the product and take all necessary measures to ensure that neither manufacturers nor importers established in the Community place products on the market which do not comply with the Directive. The risk assessment should take into account the exposure, use level, usage and skin penetration. Member States must also ensure that cosmetic products are not attributed characteristics which they do not possess.

Cosmetics packaging must bear various information about the product and the person responsible for marketing the cosmetic product within the Community, including particular precautions for use. There must also be a list of ingredients, in descending order as to the proportion they constitute of the product. Perfume and aromatic compositions must be referred to only by the word 'perfume' or 'aroma', except where these have been identified as a significant cause of allergic reactions in sensitive consumers.

These apparently straightforward requirements are made considerably more difficult to comply with by the absence of any globally harmonized nomenclature of cosmetic ingredients, despite efforts by various national trade associations. The result is that except in respect of synthetic ingredients or those subject to REACH, the information supplied in the dossier may not be totally transparent to the competent authority, and results in a cascade of questions which may or may not be readily answerable. For instance, plant nomenclature is continuously shifting as genetic techniques are brought to bear on classifications that were formerly based on similarity of structure and appearance; nor may it be possible to give a consistent response to the question of the ratio between the mass of plant required to produce a given extract and that of the extract[110] – but which certainly increase the time and cost of compliance. Further, the more information which is required to be disclosed, the lower the possibility of maintaining any useful confidentiality over formulations and supply sources. There is currently no protection for information supplied in response to regulatory requirements in cosmetics.

6.5.3 Cosmetics and REACH/CLP

Cosmetic products are exempt from some of the REACH requirements but only in relation to human health hazard exposure; other considerations, such as environmental risk or bioaccumulation may still cause a cosmetic to fall within the REACH assessment scheme.

Finished cosmetic preparations are not covered by the CLP Regulation. But where a manufacturer or importer supplies substances and mixtures, whether active ingredients or excipients, that are not yet in the finished state, this manufacturer or

---

(Product Information Requirement) for a finished cosmetic product as required by Directive 93/35/EEC (the sixth amendment to the Old Cosmetics Directive).

110. Lintner, *Cosmetic ingredients: definitions, legal requirements and an attempt to harmonize (global?) characterization* Lintner (ed.), *Global Regulatory Issues for the Cosmetics Industry*, vol. 2 (William Andrew Inc., 2009).

importer will have to classify, package and label these substances and mixtures in accordance with CLP. In addition, if these substances are placed on the market, they will also have to be notified to the C&L Inventory.[111]

The information requirements not covered by REACH are dealt with in practice through questionnaires supplied by the formulator to their suppliers to attempt to collate all of the information which could be needed either for regulatory compliance or in order to defend any potential product liability claim, since the safety of cosmetics as between producer and end user is regulated under the Product Safety Directive.[112]

6.5.4 The SCCP Opinion on Nanomaterials in Cosmetic Products

In December 2007, following on from the suggestion in the report of the Royal Society and RAE[113] that nanomaterials should be treated as new chemicals from the point of view of risk and also of skin absorption both in normal and diseased skin, the SCCP adopted an Opinion on the Safety of Nanomaterials in Cosmetic Products.[114] The specific questions addressed were whether the use of nanomaterials gave rise to any need for revisions to the *Notes for Guidance* and the Opinions previously issued in respect of nanosized titanium dioxide and zinc oxide.

The SCCP concluded that nanoparticles can be divided into two groups:

(i) soluble and/or biodegradable nanoparticles which disintegrate upon application to skin into molecular species (e.g., liposomes, microemulsions, nanoemulsions); and
(ii) insoluble and/or biopersistent particles (e.g., titanium dioxide, fullerenes, quantum dots).

For the first group, conventional risk assessment methodologies based on mass metrics *may* be adequate, whereas for the latter other metrics such as the number of particles, and their surface area as well as their size distribution are also required. It is primarily for the insoluble particles that health concerns related to possible uptake arise. Should they become systemically available, translocation/transportation and eventual accumulation in secondary target organs may occur. This feature does become important with repeated application of cosmetic products. Further, it is inevitable that insoluble nanoparticles represent a burden for the environment and a complete life cycle analysis is required. These analyses would need to be made on a case-by-case basis as there are major data gaps in the assessment of the exposure and the uptake of nanoparticles via dermal absorption, inhalation, oral ingestion and eye contact. However, if the technical

111. See s. 6.2.10 above.
112. Discussed in Ch. 3.
113. *Nanoscience and nanotechnologies: opportunities and uncertainties*, June 2004.
114. SCCP 1147/07.

specifications of nanomaterials are maintained from batch to batch, no toxicological assessment of different batches is required.

When characterizing nanoparticles for risk assessment, in general the SCCP concluded that the following properties need to be considered:

- size and shape (e.g., spherical or fibrous);
- surface area, surface charge and surface morphology;
- rheology, porosity, crystallinity and amorphocity;
- primary nanoparticles, agglomerates and/or aggregates;
- chemical composition, surface chemistry and coating;
- oxidative capacity;
- catalytic activity;
- stoichiometry (may change for large surface to volume ratios);
- dissolution kinetics and solubility;
- hydrophilicity or hydrophobicity; and
- impurities (foreign elements, chemical by-products or degradation products, etc.) and intentional or unintentional surface adsorbents (both of which determine the reactivity).

In other words, the Committee concluded that characterization of the properties of nanoparticles themselves is insufficient. The interactions of the nanoparticles within a given environment must be examined including solubility/insolubility, free radical formation, etc. But at present there is insufficient knowledge about the role of physico-chemical parameters of nanoparticles determining absorption and transport across membranes in the gut and lungs, or how once the particles are in systemic circulation these determine biokinetics and accumulation in secondary target organs. Nor are possible health effects established or *in vitro* and *in vivo* test methods validated or optimized for nanomaterials.

Although animal testing can be largely reduced for skin penetration studies, it remains essential for translocation and accumulation studies as well as for chronic toxicity studies.

As a result, for the foreseeable future each safety dossier concerning nanomaterials needs to be evaluated on a case-by-case basis.

6.5.5 The New Cosmetics Regime in Europe

It is no surprise therefore to find that in the revision of the regulatory regime for cosmetics, nanomaterials were explicitly addressed for the first time. European Regulation 1223/2009[115] (the Cosmetics Regulation), passed in 2009, will enter into force on 11 July 2013 with the exception of specific articles which will have come into force earlier, on 1 December 2010 and 11 January 2013. The new Cosmetics Regulation completely replaces the Old Cosmetics Directive, with

115. OJ L 22 Dec. 2009 342/59.

the stated objective of simplifying procedures and streamlining terminology to reduce ambiguity and administrative burden. However, the key elements of the Directive have not been changed; the definition of a cosmetic product is broad, with no intermediate category between cosmetics and pharmaceuticals. Control remains on an in-market basis by the EU Member States, as opposed to pre-market approval (save insofar as ingredients require pre-approval under REACH). The responsibility remains with the person placing a cosmetic product on the EU market, and the product must remain traceable through the supply chain. A safety evaluation based on risk rather than hazard is a key element.

The nomenclature issue is addressed for the first time, with the European Commission being charged with the compilation of a glossary of common ingredient names to ensure uniform labelling and to facilitate identification of cosmetics ingredients. The glossary is expressly *not* to constitute a limitative list of substances which can be used in cosmetic products.

The list of prohibited substances remains in effect, but in addition substances which are CMR[116] are automatically prohibited for use in cosmetic products from the date of being so classified. There is no requirement for additional listing in Annex II to the Cosmetics Regulation for the ban to become effective, rather, implementing measures will be required only to authorize for use in cosmetic products those CMR substances that meet conditions, laid down in the Cosmetics Regulation, for the derogation from the general prohibition principle.

An EU-wide two-track notification method is to be introduced under Article 13, to avoid different national notification methods. Specific information about each product must be notified prior to the placing of the product on the market: name; category; country of origin (in the case of imports from third countries); the Member State in which the product is placed on the market; any nanomaterials or CMR substances used in the product; and the framework formulation for the product.[117] The European Commission will inform all competent authorities about the notification where there are differing responsibilities for the information. For instance, the information about the framework formulation is intended solely for Poison Control Centres so that they can take quick and suitable action for treatment of poisoning incidents. The other notification data, by contrast, are only intended for the competent national authorities.

The Commission is currently preparing a template for future notifications. All notifications should be undertaken electronically and managed centrally. As soon as the system has been introduced, national notifications will become redundant. Notification is only required for those products which are still being marketed when Article 13 comes into force.

116. Classification is in accordance with categories 1A, 1B or 2 under Part 3 of Annex VI to Regulation (EC) No. 1272/2008 (GHS Regulation: Global Harmonised System of Classification and Labelling of Chemicals).
117. Article 13.

6.5.6 NANOMATERIALS UNDER THE NEW REGIME

6.5.6.1 Objectives and Definition

Nanomaterials are explicitly regulated for the first time in cosmetics legislation. The recitals to the Cosmetics Regulation reflect the SCCP Opinion on Nanomaterials, stating at 29–31 that:

> (29) The use of nanomaterials in cosmetic products may increase with the further development of technology. In order to ensure a high level of consumer protection, free movement of goods and legal certainty for manufacturers, it is necessary to develop a uniform definition for nanomaterials at international level. The Community should endeavour to reach an agreement on a definition in appropriate international fora. Should such an agreement be reached, the definition of nanomaterials in this Regulation should be adapted accordingly.
>
> (30) At present, there is inadequate information on the risks associated with nanomaterials. In order to better assess their safety the Scientific Committee on Consumer Safety (SCCS) should provide guidance in cooperation with relevant bodies on test methodologies which take into account specific characteristics of nanomaterials.
>
> (31) The Commission should regularly review the provisions on nanomaterials in the light of scientific progress.

In addition, recital 35 stipulates that:

> The SCCS should give opinions where appropriate on the safety of use of nanomaterials in cosmetic products. These opinions should be based on full information being made available by the responsible person.

The definition, in Article 2(k), is brief:

> (k) 'nanomaterial' means an insoluble or biopersistent and intentionally manufactured material with one or more external dimensions, or an internal structure, on the scale from 1 to 100 nm.

This goes further than the tentative conclusion of the SCCP that for soluble and/or biodegradable nanoparticles conventional risk assessment may be adequate, simply excluding all soluble nanoscale systems or nanoscale systems which are unstable under physiological conditions such as nanoemulsions, liposomes, nanosomes or niosomes, from the new regime. Like the definition in the draft Novel Foods Regulation, the reference to internal structures on the scale from 1 to 100 nm is intended to include materials consisting of aggregates or agglomerates of nanoparticles. The absence of reference to properties (other than solubility and persistence) or specific surface area makes this more workable from a regulatory perspective than some of the more complex and potentially subjective definitions which have been discussed, and also a more blunt instrument. However, the Cosmetics Regulation specifically recognizes the indeterminate state of the international discussion of nanomaterials, and at Article 2(3) provides power to the Commission to adapt the definition in Article 2(1)(k) to 'technical and scientific progress and to definitions subsequently agreed at international level'.

6.5.6.2 Notification and Labelling Requirements

Article 16 sets out comprehensive new requirements to be met by nanomaterials or products containing nanomaterials beginning, rather repetitively given the Regulation's overall requirement that cosmetic products be safe, with a stipulation that for every cosmetic product that contains nanomaterials, a high level of protection of human health shall be ensured. Nanomaterials used as colorants, UV-filters or preservatives regulated under Article 14 are excluded from the Article 16 regime, unless expressly specified.[118] The colorants and UV-filters regulated under Article 14 include titanium dioxide[119] but nothing in those provisions specifies whether or not the substance is covered in bulk as well as nanoform, or only one or the other.

Article 16 requires an additional notification to be made where a cosmetic product containing nanomaterials is to be placed on the market. The responsible person (manufacturer, importer or distributor[120]) must notify[121] the Commission by electronic means six months before placing the product on the market, except where the product has already been placed on the market by the same responsible person before 11 January 2013. If so, then the notification can be made between 11 January 2013 and 11 July 2013. The notification must contain:

(a) the identification of the nanomaterial including its chemical name (IUPAC: International Union of Pure and Applied Chemistry) and other descriptors as specified in point 2 of the Preamble to Annexes II to VI;
(b) the specification of the nanomaterial including size of particles, physical and chemical properties;
(c) an estimate of the quantity of nanomaterial contained in cosmetic products intended to be placed on the market per year;
(d) the toxicological profile of the nanomaterial;
(e) the safety data of the nanomaterial relating to the category of cosmetic product, as used in such products; and
(f) the reasonably foreseeable exposure conditions.[122]

118. Article 16(2) Cosmetics Regulation.
119. Annex IV 143; Annex VI 27.
120. Article 5(4)–(6).
121. Or may designate another legal or natural person in writing for the notification of nanomaterials.
122. Article 16(3).

If the Commission has concerns regarding the safety of a nanomaterial, either when it is notified or at any later time,[123] the Commission must request the SCCS to give its opinion within six months on the safety of such nanomaterial for use in the relevant categories of cosmetic products and on the reasonably foreseeable exposure conditions. The fact of such a request will be public. Where the SCCS finds that any necessary data is lacking, the Commission shall request the responsible person to provide such data within an explicitly stated reasonable time. Notably, there is no provision for extension of this time period which may make it advisable to consult the Commission or SCCS in advance of notifying, in case time is needed to develop or validate suitable test methods for providing the additional data which may be required. Once the additional data has been submitted, the SCCS has a further six months to deliver its opinion, which will also be public. Once again, commercial interests in confidentiality of information take second place to the imperative for transparency in view of the public concerns over nanotechnology.

The opinion or an insufficiency of data may lead the Commission to add the nanomaterial to Annex II (prohibited materials) or III (materials which can be used only subject to restrictions) in view of any risks to health the nanomaterial may pose.

Article 19(1)(g) requires nanomaterials to be separately labelled in the list of ingredients. To this end, all INCI designations of nanomaterials contained in a product must bear the suffix '(Nano)', as for instance 'Titanium Dioxide (Nano)'.

6.5.6.3 Public inventory

The Commission is to make available by 11 January 2014 a catalogue of all nanomaterials used in cosmetic products placed on the market, including (in a separate section) those used as colorants, UV-filters and preservatives, indicating the categories of cosmetic products and the reasonably foreseeable exposure conditions.[124] This catalogue shall be regularly updated thereafter and be made publicly available. It will also submit an annual status report on the use of nanomaterials in cosmetics to the European Parliament and Council, including summarizing the progress made in developing nano-specific assessment methods and safety assessment guides, and information on international cooperation programmes.

Article 20 requires the European Commission also to draw up a report on the use of advertising claims for cosmetic products by 11 July 2016. To this end, the European Commission, together with the Member States, is to initially define criteria that could be used as the basis for claims about cosmetic products. In this context the basic requirements of advertising claims apply, that is, they

123. For example, if information provided by a third party raises grounds for later concern – Art. 16(5).
124. Article 16(10)(a).

may not, of course, mislead the consumer. Hence Article 20(1) of the Cosmetics Regulation prohibits the use of text, names, trademarks, pictures and figurative or other signs in the labelling, sale and advertising of cosmetic products to imply that these products have characteristics or functions which they do not have. This will prevent the inclusion of the 'nano' designation in products which do not in fact contain any nanomaterials.

The whole regime is to be kept under review in the light of scientific progress and amended where necessary, with the first review to be undertaken by 11 July 2018.[125]

6.6 ENVIRONMENT AND WASTE

6.6.1 Overview

In 2003, an integrated strategy on the environment and health (the European Strategy) was introduced[126] which formed the basis for the European Health Action Plan 2006–2010.[127] The European Strategy mentioned nanotechnology in the context of other ongoing Commission work or research, and the need to better integrate Environment and Health into nanotechnology policy. The mid term review of the European Health Action Plan[128] highlighted nanotechnology as an emerging issue.

The Commission's Gap Analysis identified a number of environment and waste directives as potentially being capable of application to control the impact of nanomaterials. It highlighted the role of the Directives on Integrated Pollution Prevention and Control (IPPC), the control of major-accident hazards involving dangerous substances (Seveso II), the framework for Community action in the field of water policy[129] (the Water Framework Directive) and a number of waste directives.

The European Parliament accepted that these were among the provisions which should be reviewed but added in those on air quality, and noted the need in respect of this, water quality and waste in particular to supplement the current mass-based measurements by metrics based on particle number and/or surface area to adequately address nanomaterials. The Parliament was concerned that waste

125. Article 16(11).
126. Communication from the Commission to the Council, the European Parliament and the European Economic and Social Committee – A European Environment and Health Strategy (COM/2003/0338 final).
127. Communication of 9 Jun. 2004 from the Commission: 'The European Environment & Health Action Plan 2004–2010' COM(2004) 416 – OJ C 49, 28 Feb. 2006.
128. Communication of 11 Jun. 2007 from the Commission: 'Mid Term Review of the European Environment and Health Action Plan 2004–2010' COM(2007) 314 final – OJ C 191, 17 Aug. 2007.
129. Directive 2000/60/EC of the European Parliament and of the Council of 23 Oct. 2000 establishing a framework for Community action in the field of water policy OJ L 327, 22 Dec. 2000.

legislation in the absence of nano-specific provisions may not apply correctly and the Parliament's 2009 Resolution called specifically on the Commission to evaluate the need to review waste legislation concerning *inter alia*:

– a separate entry for nanomaterials in the list of waste established by Decision 2000/532/EC;[130]
– a revision of the waste acceptance criteria in landfills in Decision 2003/33/EC;[131]
– a revision of relevant emission limit values for waste incineration to supplement the mass-based measurements by metrics based on particle number and/or surface.

It further called on the Commission to evaluate the need to review legislation to address nanomaterials that are created as unintended by-products of combustion processes in a cost-effective manner.

6.6.2 Integrated Pollution Prevention and Control

The IPPC Directive[132] requires industrial and agricultural activities with a high pollution potential to have a permit. This permit can only be issued if certain environmental conditions are met, so that the companies themselves bear responsibility for preventing and reducing any pollution they may cause.

What constitutes a high pollution potential is defined in Annex I to the Directive (broadly, the energy industries, production and processing of metals, mineral industry, chemical industry, waste management, livestock farming, etc.). Approximately 52,000 industrial installations around the EU are covered.

In order to receive a permit an industrial or agricultural installation must comply with certain basic obligations. As relevant to nanomaterials, these include:

– use the best available pollution-prevention measures which produce the least waste, use less hazardous substances and enable the substances generated to be recovered and recycled;
– prevent all large-scale pollution;
– prevent, recycle or dispose of waste in the least polluting way possible;
– ensure accident prevention and damage limitation;
– return sites to their original state when the activity is over.

130. Commission Decision 2000/532/EC of 3 May 2000 replacing Decision 94/3/EC establishing a list of wastes pursuant to Art. 1(a) of Council Directive 75/442/EEC on waste and Council Decision 94/904/EC establishing a list of hazardous waste pursuant to Art. 1(4) of Council Directive 91/689/EEC on hazardous waste (OJ L 226, 6 Sep. 2000, 3).
131. Council Decision 2003/33/EC of 19 Dec. 2002 establishing criteria and procedures for the acceptance of waste at landfills pursuant to Art. 16 of and Annex II to Directive 1999/31/EC (OJ L 11, 16 Jan. 2003, 27).
132. OJ L 24, 29 Jan. 2008, 8.

In addition, the permit is subject to a number of potential specific requirements, including:

- emission limit values for polluting substances;
- any soil, water and air protection and waste management measures required;
- measures to be taken in case of leaks, malfunctions, temporary or permanent stoppages;
- minimization of long-distance or transboundary pollution; and
- release monitoring.

Permits, issued by the competent authority of the Member State concerned, must include:

- a description of the installation, its site conditions and the nature and scale of its activities;
- the materials and substances used or generated;
- the sources of emissions from the installation, and the nature and quantities of foreseeable emissions into each medium, as well as their effects on the environment;
- the proposed technology and other techniques for preventing or reducing emissions;
- measures for monitoring emissions and preventing and recovering waste; and
- possible alternative solutions.

Within the limits of commercial and industrial secrecy, this information must be made available to the public and, if the project is likely to have cross-border effects, the other Member States. Each Member State must submit this information to interested parties in its territory so that they can give their opinion. The decision to license or reject a project, the arguments on which this decision is based and possible measures to reduce the negative impact of the project are also made public and sent to the other Member States concerned.

Each Member States is responsible for inspecting industrial installations and ensuring they comply with the Directive. Further, the Commission, the Member States and the industries concerned regularly exchange information on best available techniques.

The Commission considers that in principle, the IPPC Directive could be used to control environmental impacts of nanomaterials and nanomaterials issues at IPPC installations through the inclusion of such considerations into the Commission's Best Available Technology Reference Document process. However, this would require the inclusion of nano-engineered materials in the list (at Annex I of the Directive) of industrial activities with a high pollution potential. Chemical installations for the production of a wide-ranging list of organic and inorganic chemicals are listed, but there are no specific references to synthesized or manufactured nanoparticles. It is unlikely therefore that the IPPC Directive would apply at present.

6.6.3 Major Accidents

The Seveso II Directive[133] applies to establishments where named dangerous substances or substances falling within certain classification categories are present above specific quantities (or thresholds). It is applicable to any establishment where dangerous substances are present, or likely to be produced as a result of an accident, in quantities equal to or in excess of the quantities listed in the Annex. The list of named substances in the Annex is accompanied by a list of categories of substances, which in practice broadens the scope. It does not apply to the carriage of dangerous substances by road, rail, air and inland waterways or to waste landfill sites, which are the subject of separate legislation.

It imposes a general obligation on operators to take all measures necessary to prevent major accidents and to limit their consequences for man and the environment. The Commission's Gap Analysis observes that if certain nanomaterials are found to demonstrate a major-accident hazard, they may be categorized, together with appropriate thresholds, in the context of the Directive. At present, however, they are not covered.

6.6.4 Water

The Water Framework Directive sets common principles and an overall framework for action to improve the aquatic environment. Management plans were produced for each river basin district within the EU, initially for the period 2009–2015. They must be implemented in 2012 and aim to prevent deterioration, enhance and restore bodies of surface water, achieve good chemical and ecological status of such water by 2015 at the latest and to reduce pollution from discharges and emissions of hazardous substances. The management plans for river basin districts can be complemented by more detailed management programmes and plans for a sub-basin, a sector or a particular type of water.

The Directive aims progressively to reduce pollution from 'priority substances' and to phase out emissions, discharges and losses of priority hazardous substances to water. Priority substances are those included in a list, established in 2001. The Commission's Gap Analysis points out that nanomaterials could be included among the Priority Substances depending on their hazardous properties. If so, Environment Quality Standards for these materials would be proposed by the Commission. Member States are required to establish quality standards for pollutants representing a risk to groundwater, which could also include nanomaterials. However, nanomaterials are not currently listed as priority substances and temporary deterioration of bodies of water is not in breach of the requirements of this Directive if it is the result of circumstances which are exceptional or could

133. Directive 96/82/EC on the control of major-accident hazards involving dangerous substances (OJ L 10, 14 Jan. 1997, 13–33) as amended by Directive 2003/105/EC of the European Parliament and of the Council of 16 Dec. 2003 (OJ L 345, 31 Dec. 2003, 97–105).

not reasonably have been foreseen and which are due to an accident, natural causes or force majeure. Accordingly, not only does the Directive not presently cover nanomaterials but there is also no breach if the pollution is temporary, exceptional and accidental. It is, therefore, not likely to be as effective in controlling pollution by nanomaterials as the IPPC Directive.

6.6.5 Waste

6.6.5.1 Waste Framework

The general framework on waste, defined as any substance or object which the holder discards or intends or is required to discard, is set by Directive 2008/98/EC[134] which imposes an obligation on Member States to ensure that waste treatment does not adversely affect health and the environment. It applies to most forms of waste in which nanomaterials are likely to be present. In general, nanoparticles are likely to follow the material or the substance in which they are contained and their fate depends on the way these wastes are treated.

6.6.5.2 Landfill

Wastes are typically disposed of in landfills or incinerated. Very little is known about the long-term behaviour of nanoparticles in a landfill. Release depends on the nanoparticles' mobility as well as on the degradability of the host material for fixed particles. There are no provisions for waste recovery or disposal options relating to the definition, characteristics and liability for waste containing nanomaterials.

As waste is not considered to be a substance, mixture or article under CLP, waste treatment operators are not considered as downstream users. At the same time waste treatment operators will not receive SDS on how to handle a substance or mixture during the waste phase. Similarly, as long as residues from waste treatment operations are waste, that is, they are disposed of (e.g., land-filled), they do not fall under the scope of CLP. Residues which are recovered as substances or mixtures do, however, fall under the scope of CLP.

Any establishment or undertaking intending to carry out waste treatment must obtain a permit from the competent authorities who determine notably the quantity and type of treated waste, the method used as well as monitoring and control operations. This in principle allows the Member States to set conditions for the treatment of nanomaterials as part of general waste processing, including landfill under Directive 1999/31/EC.[135] Dangerous waste must be stored and treated in conditions that ensure the protection of health and the environment. They must not,

134. OJ L 312 22 Jan. 2008, 3.
135. OJ L 182, 16 Jul. 1999, 1–19.

in any case, be mixed with other dangerous waste and must be packaged or labelled in line with international or Community regulations.

6.6.5.3 Hazardous Waste

Wastes defined as hazardous are treated separately under the Hazardous Waste Directive,[136] Annex III of which defines which wastes are hazardous and sets up the European Waste List. Wastes containing nanomaterials could be classified as hazardous, if the nanomaterial displays relevant properties which render the waste hazardous. These include:

- 'Irritant', meaning non-corrosive substances and preparations which, through immediate, prolonged or repeated contact with the skin or mucous membrane, can cause inflammation.
- 'Harmful', meaning substances and preparations which, if they are inhaled or ingested or if they penetrate the skin, may involve limited health risks.
- 'Toxic', meaning substances and preparations (including very toxic substances and preparations) which, if they are inhaled or ingested or if they penetrate the skin, may involve serious, acute or chronic health risks and even death.
- 'Ecotoxic' meaning substances and preparations which present or may present immediate or delayed risks for one or more sectors of the environment.

If a Member State believes that any waste should be classified as hazardous as a result of the presence of nanomaterials leading to these properties, it can notify the Commission which will consider whether such waste should be added to the European Waste List.[137] To date, no reclassification has taken place and accordingly nanomaterial-bearing wastes are not regulated by this Directive.

6.6.5.4 Restriction on Use of Hazardous Substances

Legislation has also been adopted to deal with particular waste streams or specific waste treatment processes, such as incineration. Of these, the Directive on the Restriction on use of Hazardous Substances (ROHS Directive)[138] has now been recast and includes reference to nanotechnology for the first time. The others remain silent; the Commission's Gap Analysis comments that if the need for more specific provisions is established, appropriate action can be proposed or implemented under the current legislative framework.

---

136. Directive 91/689/EEC; OJ L 377, 31 Dec. 1991.
137. Article 4 Hazardous Waste Directive.
138. Directive 2011/65/EU of the European Parliament and of the Council of 8 Jun. 2011 on the Restriction of the Use of Certain Hazardous Substances in Electrical and Electronic Equipment OJ L 174/88, 1 Jul. 2011.

The ROHS Directive requires that new electrical equipment put on the market does not contain more than the permissible maximum concentration values of hazardous substances. Hazardous substances for this purpose are those listed in Annex II, which currently does not list any nanomaterials. A regulatory gap was identified as arising from the fact that these thresholds were set on the basis of the bulk forms of the relevant hazardous substances, which may not have been appropriate for the same substances in nanoform. Accordingly, although calls for nanosilver and long MWCNTs to be banned were not followed, the situation is being kept under review as the recast directive recites that:

> (16) As soon as scientific evidence is available, and taking into account the precautionary principle, the restriction of other hazardous substances, including any substances of very small size or with a very small internal or surface structure (nanomaterials) which may be hazardous due to properties relating to their size or structure, and their substitution by more environmentally friendly alternatives which ensure at least the same level of protection of consumers should be examined. To this end, the review and amendment of the list of restricted substances in Annex II should be coherent, maximise synergies with, and reflect the complementary nature of the work carried out under other Union legislation, and in particular under Regulation (EC) No 1907/2006 [REACH] while ensuring the mutually independent operation of this Directive and that Regulation. Consultation with the relevant stakeholders should be carried out and specific account should be taken of the potential impact on SMEs.

At the time of writing, negotiations are ongoing as to a recast of the related Directive 2002/96/EC on waste electrical and electronic equipment[139] which, like the original form of the ROHS Directive, currently contains no reference to nanomaterials. A Common Position has been reached on the first reading, but to date there is no proposal to introduce any mention of nanotechnology.

6.6.6 Air

As to air quality, other than measures specific to emitters such as large-scale combustion plants, this is governed by Directive 2008/50/EC on ambient air quality and cleaner air for Europe[140] (the Air Quality Directive).

This Directive lays down measures aimed at defining and establishing objectives for ambient air quality designed to reduce harmful effects on health and the environment; assessing the ambient air quality in Member States on the basis of common methods and criteria; and collating information on ambient air quality in order to monitor long-term trends.

---

139. OJ L 37/24 13 Feb. 2003.
140. OJ L 152, 11 Jun. 08 P. 0001–0044.

It is not, therefore, aimed at affecting emissions so much as knowing what is in the air at what level. Further, it focuses on stipulated pollutants: sulphur dioxide, nitrogen dioxide and oxides of nitrogen, particulate matter (PM10 and PM2.5[141]), lead, benzene and carbon monoxide as well as ozone. There is no reference to ultrafine particles,[142] the category within which nanoparticles would naturally fall. Arguably, therefore, it does not cover them at all. The Directive sets thresholds for assessment for each pollutant, criteria for the assessment method (in particular the siting of sampling points), reference methods for measurement, limit values for the protection of human health and the environment, the target and the obligation of reducing exposure for the population to PM 2.5, information thresholds and alert thresholds, critical levels for the protection of vegetation and the list of information to be included in action plans for improvement in air quality.

The controversy over measurement techniques for nanoparticles, let alone what threshold levels would be appropriate, means that it is unlikely that these will be added to the list of pollutants, or the focus of Member States' measurement and analysis efforts in the near future.

6.7 OCCUPATIONAL HEALTH AND SAFETY

Occupational health and safety legislation in the EU comes under the Framework Directive 89/391 on the introduction of measures to encourage improvements in the safety and health of workers at work (the OHS Framework Directive).[143] It lays down general principles as to the protection of workers against occupational accidents and diseases: the prevention of risks, the protection of safety and health, the assessment of risks, the elimination of risks and accident factors and the training of workers. The prevention of risks requires, amongst other things, evaluating risks and combating them at source; adapting to technical progress; and replacing the dangerous by the non- or the less dangerous. This implies taking note of new information around the risks related to particular substances as it becomes available and, where possible, acting upon it to reduce any risk to workers arising from it.

The employer's responsibilities include evaluating all risks to the safety and health of workers, including in the choice of work equipment, the chemical substances or preparations used, and the fitting-out of work places. A series of 'daughter directives' have been passed over the years introducing exposure limits for workers to a large number of chemical substances, but none of these mention nanoparticles or nanomaterials.[144] The OHS Framework Directive principles are

141. Particles less than 10 µm or 2.5 µm in size, respectively.
142. Defined as particles less than 0.1 µm (100 nm) in size.
143. OJ L 393 of 30 Dec. 1989.
144. Directive 2009/161/EU of 17 Dec. 2009 establishing a third list of indicative occupational exposure limit values in implementation of Council Directive 98/24/EC and amending Commission Directive 2000/39/EC; Directive 2009/148/EC of 30 Nov. 2009 on the protection of workers from the risks related to exposure to asbestos at work; Directive 2006/15/EC of 7

therefore those which apply to laboratories or manufacturing premises where nanomaterials are made or used.

This area might be expected to be the focus of significant legislative activity given that in 2009 two Chinese workers were reported to have died from exposure to nanoparticles.[145] However, the editors of the journal which published the report later noted trenchant criticisms of the report – one commentator suggested that the title should have been 'Poor working conditions cause pleural effusion, pulmonary fibrosis and granuloma'[146] – and accepted that

> There is no formal proof that nanoparticle exposure at this workplace caused pulmonary disease in and the death of several workers in a primitive workplace without any occupational safety measures. However, Song et al. excluded a number of fateful differential diagnoses and came to the conclusion that nanoparticles were the most likely cause.
>
> Conclusions by analogy were made from what is known about quartz, asbestos and air pollution.[147]

The Commission's Gap Analysis was drawn up before this report was published, and concluded that the OHS Framework Directive fully applies to nanomaterials, so that it is incumbent on employers to carry out a risk assessment and, where a risk is identified, take measures to eliminate this risk. Further, the Directive requires consultation with workers' or their representatives where new technologies are introduced, as regards the working conditions and the working environment.[148]

It is of course not yet possible to lay down appropriate rules regarding exposure to nanomaterials since there are no exposure scenarios fully worked out which could be used as the basis. The employer is, therefore, left to work out its own arrangements under its general responsibility for workers' health and safety. Given that the risks are not fully understood, provision of personal safety equipment is likely to be necessary on a precautionary basis to prevent risks to many workers

---

Feb. 2006 establishing a second list of indicative occupational exposure limit values in implementation of Council Directive 98/24/EC and amending Directives 91/322/EEC and 2000/39/EC; Directive 2004/37/EC of 29 Apr. 2004 on the protection of workers from the risks related to exposure to carcinogens or mutagens at work (Sixth individual Directive within the meaning of Art. 16(1) Directive 89/391/EEC); Directive 200/39/EC of 8 Jun. 2000 establishing a first list of indicative occupational exposure limit values in implementation of Council Directive 98/24/EC on the protection of the health and safety of workers from the risks related to chemical agents at work; Directive 98/24/EC of 7 Apr. 1998 on the protection of the health and safety of workers from the risks related to chemical agents at work (14th individual Directive within the meaning of Art. 16(1) of Directive 89/391/EEC); and Directive 91/322/EEC of 29 May 1991 on establishing indicative limit values by implementing Council Directive 80/1107/EEC on the protection of workers from the risks related to exposure to chemical, physical and biological agents at work.

145. Song, Li and Du. *Exposure to Nanoparticles Is Related to Pleural Effusion, Pulmonary Fibrosis and Granuloma*. Eur Respir J 34 (2009): 559–567.
146. *To the editors* Eur Respir J January 35, no. 1 (2010): 226.
147. Editorial, Eur Respir J January 35, no. 1 (2010): 227.
148. Articles 11 and 12 of the Framework Directive 89/391/EEC.

handling nanomaterials. This certainly appears to be the conclusion from work at the OECD and the US National Institute on Occupational Safety and Health.[149]

The use of personal protective equipment by workers in the EU is subject to minimum requirements.[150] Such equipment must be appropriate for the risks involved, without itself leading to any increased risk; correspond to existing conditions at the workplace; take account of ergonomic requirements and the worker's state of health; and fit the wearer correctly after any necessary adjustment. Further, if the worker needs to wear more than one such piece of equipment simultaneously, then the equipment must be compatible. The employer must provide the appropriate equipment free of charge and he must ensure that it is in good working order and hygienic condition. The employer must also organize training and demonstration on the use of the equipment.

The adequacy of current European legislation to protect workers from risks posed by exposure to nanomaterials at work was questioned in the European Parliament's 2009 Resolution, which calls specifically on the Commission to evaluate the need to review worker protection legislation, highlighting the issues of:

- the use of nanomaterials only in closed systems or in other ways that exclude exposure of workers as long as it is not possible to reliably detect and control exposure;
- a clear assignment of liability to producers and employers arising from the use of nanomaterials; and
- whether all exposure routes (inhalation, dermal and other) are addressed.

Curiously, although a literature review was published in 2009[151] there are no guidelines available through the European Agency for Safety and Health at Work dealing with exposure to nanoparticles. It remains to be seen when and how any additional legislation or guidance will be introduced.

149. Discussed in Chs 5 and 7 respectively.
150. Directive 89/656/EC on the minimum health and safety requirements for the use by workers of personal protective equipment at the workplace OJ L 393, 30 Dec. 1989, 18–28.
151. Kaluza et al., *Workplace Exposure to Nanoparticles*.

Chapter 7

Beyond Europe: An Overview

As is apparent from the range and number of participants involved in the various discussions at international level, many countries are grappling with the questions around whether and if so how best to regulate nanotechnology. In this chapter, a necessarily brief overview of the approaches being taken in selected sectors – food, chemicals, environment and occupational health and safety – in some major jurisdictions outside the EU is set out.

7.1 UNITED STATES

7.1.1 Overview

The United States is a front-runner in nanotechnology research and development, with the largest number of patents and patent applications relating to nanotechnologies of any country in the world. The federally funded interagency National Nanotechnology Initiative, launched in 2000, was one of the earliest national research programmes established to coordinate national efforts in the field. Participating agencies have developed three 'signature' programmes through interagency collaboration: nanoelectronics for 2020 and beyond; sustainable manufacturing: creating the industries of the future; and nanotechnology for solar energy collection and conversion.

In the field of regulation, however, the United States has been slow to institute any new regulation specifically directed at nanotechnology or nanomaterials, relying instead on interpretation of existing regulations to address issues as they arise.

7.1.2 Food and Drugs

7.1.2.1 FDA Nanotechnology Task Force

The Food and Drug Administration (FDA) regulates products under both the Food, Drug and Cosmetics Act and the Public Health Service Act. Its Nanotechnology Task Force was formed in August 2006 to develop regulatory approaches that would encourage the continued development of innovative, safe and effective FDA-regulated products that use nanotechnology materials. The agency expects that many of the nanotechnology products it will regulate will span the regulatory boundaries between drugs, medical devices and biologics. These, then, would be regulated under the rules established for 'combination products'.

The Task Force reported in 2007. In line with the problem of inadequate scientific knowledge identified in several international and regional projects, its initial recommendations related to improving scientific knowledge of nanotechnology to help ensure the agency's regulatory effectiveness, particularly with regard to products not subject to pre-market authorization requirements. The report also addressed the need to evaluate whether the tools available to describe and evaluate nanoscale materials were sufficient, and the development of additional tools where necessary.

The Task Force also assessed whether the FDA can identify products containing nanoscale materials, the scope of FDA's authorities to evaluate the safety and effectiveness of such products, whether FDA should require or permit products to be labelled as containing nanoscale materials, and whether the use of nanoscale materials in FDA-regulated products raises any issues under the National Environmental Policy Act.

The Task Force concluded that the agency's authorities are generally comprehensive for products subject to pre-market authorization requirements, such as drugs, biological products, devices, and food and colour additives. The FDA can obtain from manufacturers the detailed scientific information it needs to review the safety and, as appropriate, effectiveness of products.

For products not subject to pre-market authorization requirements, such as dietary supplements, cosmetics, and food ingredients that are generally recognized as safe (GRAS), manufacturers are generally not required to submit data to FDA prior to marketing, and the agency's oversight capacity is accordingly less comprehensive. Limited information need be provided where a food additive is generally recognized as safe for instance – but it is not clear whether a nanoscale version of a GRAS additive is still GRAS. Nor is there any information as to what data should be submitted for approval of a nanoscale food additive. The FDA can take action if a new version of an existing approved or GRAS additive raises safety concerns, but this will depend on feedback from the marketplace.

The Task Force accordingly made various recommendations to address the regulatory challenges that may be presented by such products. A number of recommendations dealt with requesting data and other information about the effects of nanoscale materials on safety and, as appropriate, effectiveness of products. Other recommendations were that the FDA could provide guidance to manufacturers about when the use of nanoscale ingredients may require submission of additional data, change the product's regulatory status or pathway, or merit taking additional or special steps to address potential safety or product quality issues. The Task Force recommended seeking public input on the adequacy of the FDA's policies and procedures for products that combine drugs, biological products, and/or devices containing nanoscale materials to serve multiple uses, such as both a diagnostic and a therapeutic intended use. Finally, the Task Force recommended encouraging manufacturers to communicate with the agency early in the development process for products using nanoscale materials, particularly with regard to highly integrated combination products.

7.1.2.2 Regulatory Response

The FDA accordingly began asking for data on size in a routine guidance update for food contact materials in December 2007,[1] asking sponsors to describe particle size, size distribution, and morphology, as well as any size-dependent properties if particle size is important to obtaining the technical effect or may relate to toxicity. In March 2009, guidance on technical data to be submitted in a Food Additive Petition[2] was updated to state (similarly) that if the particle size is important for the additive to achieve its intended technical effect, such that the additive is produced or processed using techniques or tools that manipulate the particle size and may contain altered particles that are formed as manufacturing by-products, then additional data should be included as appropriate. This should address the size (average and distribution), shape, surface area (average and distribution), surface charge (zeta potential), and morphology of the particles, as well as any other size-dependent properties (e.g., agglomeration, aggregation, dispersion).

The same approach is expected to be taken with other guidance documents when they are updated. The toxicology guidances, both for indirect additives and direct additives, were in late 2010 under consideration for revision. The revised toxicology guidance for direct additives will include requirements for GRAS substances – although greater uncertainties in regard to toxicity of nanoscale versions of otherwise GRAS substances may make it difficult or impossible to support a GRAS determination unless there is sufficient validated toxicity information on the nanoscale version of the compound.[3]

---

1. Section II.A.5. Physical/Chemical Specifications, Guidance for Industry: Preparation of Premarket Submissions for Food Contact Substances: Chemistry Recommendations, available at <www.fda.gov/Food/GuidanceComplianceRegulatoryInformation/GuidanceDocuments/FoodIngredientsandPackaging/ucm081818.htm>.
2. Guidance for Industry: Recommendations for Submission of Chemical and Technological Data for Direct Food Additive Petitions, available at <www.fda.gov/Food/GuidanceComplianceRegulatoryInformation/GuidanceDocuments/FoodIngredientsandPackaging/ucm124917.htm>.
3. Acting Director of the Office of Food Additive Safety quoted in *FDA's Mitchell Cheeseman Discusses Regulatory Challenges and Changes for Food Contact Materials at Keller & Heckman Seminar* on <www.packaginglaw.com>.

The Task Force continues to identify and recommend ways to address any knowledge or policy gaps that exist so as to better enable the agency to evaluate possible adverse health effects from FDA-regulated products that use nanotechnology materials. In its 2011 Strategic Plan[4] the FDA announced that it has established Nanotechnology Core Centers to provide critically needed equipment and technical staff to conduct medical product assessment and safety research in support of regulatory decision-making, including establishing methods to assess quality and effectiveness of products that use nanomaterials, and identify standards to be incorporated in the preclinical safety assessment of products that contain nanomaterials.

7.1.3 Chemicals

7.1.3.1 The EPA's Role

The Environment Protection Agency (EPA) is responsible for administering the Toxic Substances Control Act (TSCA) and the Inventory Update Reporting (IUR) programme, which requires manufactures of chemical substances included on the TSCA Chemical Substance Inventory[5] to periodically collect screening-level, exposure-related information on chemical substances being manufactured in or imported into the United States and to make that information available for use by the EPA and, to the extent possible, to the public. The threshold for reporting is 25,000 pounds (just over 11 tonnes) manufactured or imported at a single site. The IUR data are used to support risk screening, assessment, priority setting and management activities and constitute the most comprehensive source of basic screening-level, exposure-related information on chemicals available to the EPA. These reports are made every four years, with the most recent being in September 2011.

7.1.3.2 The Nanoscale Materials Stewardship Program

In view of the inadequacy of the IUR to capture information on a rapidly evolving field in a timely manner, the EPA ran a voluntary US Nanoscale Materials Stewardship Program (NMSP) from January 2008 to December 2009 to encourage submission and development of information for nanoscale materials, to help provide a firmer scientific foundation for regulatory decisions. It had two parts: a Basic Program for reporting available information on engineered nanoscale materials that were manufactured, imported, processed or used; and an In-Depth Program to voluntarily develop data, including testing, over a longer time frame.

4. Advancing Regulatory Science at FDA: A strategic plan, August 2011.
5. A list of each chemical substance which is manufactured or processed in the United States, which the EPA is required to keep by s. 8(b) of TSCA.

While 31 organizations participated in the Basic Program, only four participated in the In-Depth Program.

The Agency released its Interim Report on the NMSP in January 2009; a final report was due in 2010 but did not appear yet to be available at the time of writing.

Although the NMSP provided the EPA with information regarding a limited number of nanoscale materials in commerce, a significant number of the environmental health and safety data gaps remain. The EPA is therefore taking a number of regulatory actions using its powers under the TSCA, since many nanoscale materials are regarded as 'chemical substances' under the definition in the TSCA, which includes any organic or inorganic substance of a particular molecular identity, including any combination of such substances occurring in whole or in part as a result of a chemical reaction or occurring in nature, any chemical element or uncombined radical, unless excluded.[6]

The EPA's regulatory options include premanufacture notifications; Significant New Use Rules (SNURs); an information gathering rule; and a test rule, all discussed further below. Further, if the EPA determines that a new chemical will present unreasonable risk, it can either limit the amount or impose other restrictions on the substance via an immediately effective proposed rule, or completely prohibit the substance by issuing a proposed order or applying to a US District Court for an injunction.[7] The EPA can also allowed the manufacture of new chemical nanoscale materials under the terms of certain regulatory exemptions,[8] but has only done so in circumstances where exposures were tightly controlled to protect against unreasonable risks (using, for example, protective equipment and environmental release limitations as discussed below).

These instruments enable the EPA to regulate nanoscale materials which it considers new chemical substances.[9] The EPA *TSCA Inventory Status of Nanoscale Substances – General Approach*'[10] reviews whether manufactured nanomaterials are 'new chemical substances' or 'existing chemical substances', and concludes that the same judgment standards as applied in the past (where molecular identity serves as the standard, and physical properties such as size are not taken into consideration) will be applied on a case-by-case basis. MWCNTs, single-walled CNTs, fullerenes and so on, which are carbon allotropes, are considered to be 'new' chemical substances, but nanosized titanium dioxide and silver nanoparticles, etc. will be considered 'existing' chemical substances.

---

6. 'Chemical substance' does not include: (1) any mixture; (2) any pesticide when manufactured, processed, or distributed in commerce for use as a pesticide; (3) tobacco or any tobacco product; (4) any source material, special nuclear material, or by-product material; (5) any pistol, firearm, revolver, shells, or cartridges; or (6) any food, food additive, drug, cosmetic, or device, when manufactured, processed, or distributed in commerce for use as a food, food additive, drug, cosmetic, or device. 40 C.F.R. ß 720.3(e). The exclusions are essentially substances regulated under other legislation.
7. TSCA s. 5(f) Actions.
8. Under TSCA s. 5(h)(4).
9. Ordinarily, any chemical substance not already on the TSCA Chemical Substance Inventory.
10. At <www.epa.gov/opptintr/nano/nmsp-inventorypaper2008.pdf>.

In addition to information obtained from manufacturers' submissions, the EPA has a Nanomaterial Testing Strategy. This is to conduct focused research to inform nanomaterial safety decisions that may be made under the various environmental statutes for which the EPA is responsible. It may also be useful in areas beyond the EPA's responsibilities. The strategy focuses on identifying sources, fate, transport, and exposure; understanding human health and ecological effects to inform risk assessments and test methods; developing risk assessment approaches; and preventing and mitigating risks.

7.1.3.3 Premanufacture Notifications

Anyone who plans to manufacture or import a new chemical substance for a non-exempt commercial purpose is required by section 5 of TSCA to provide the EPA with notice before initiating the activity. This premanufacture notice (PMN), must be submitted at least 90 days prior to the manufacture or import of the chemical. The EPA can take action to ensure that those chemicals that pose an unreasonable risk to human health or the environment are effectively controlled. Almost 90% of PMNs submitted to the EPA complete the review process without being restricted or regulated in any way. However, the EPA may issue a TSCA section 5(e) Order to prohibit or limit activities associated with the substance, if the EPA considers that:

(1) there is insufficient information to evaluate the human health and environmental effects of the substance; and
(2) that the substance may present an unreasonable risk of injury to human health or the environment (the 'risk-based' finding); or
(3) that the substance will be produced in substantial quantities and may be anticipated to enter the environment in substantial quantities or there may be significant or substantial human exposure (the 'exposure-based' finding).

In practice, section 5(e) Orders are almost always issued as Consent Orders that are signed by both the EPA and the chemical manufacturer. Since such orders are only issued where there is insufficient information, most section 5(e) Orders require the manufacturer or importer to develop and submit to the EPA certain toxicity or fate tests before exceeding a specified production volume. Exposure-based section 5(e) Orders consist primarily of a requirement to conduct triggered testing (plus record-keeping and 'risk notification' if the test data indicates a risk.) Risk-based TSCA section 5(e) Orders, depending on the type of concerns identified, typically also require exposure controls for people handling the substance such as gloves, goggles, respirators, specified disposal technologies or restrictions on releases to water, and hazard communication such as material safety data sheets (MSDS), labels, and training.

Since 2005, the EPA has received and reviewed over 100 new chemical notices under TSCA for nanoscale materials. It has issued several Consent Orders regulating new chemical submissions of CNTs under TSCA.[11] These may require the manufacturer to deliver samples to the EPA with a copy of the material safety data sheet for the product; conduct a 90-day inhalation toxicity study in rats with a post-exposure observation period of up to three months, including bronchoalveolar lavage fluid analysis; submit material characterization data within six months; ensure employees use gloves impervious to nanoscale particles and chemical protective clothing; and ensure employees use a US National Institute for Occupational Safety and Health (NIOSH) approved full-face respirator with an N-100 cartridge while exposed by inhalation in the work area.

The material characterization requirement of the consent order may require the manufacturer to submit, for example:

- type of multi-walled carbon nanotube (concentric cylinders or scrolled tubes; number of walls/tubes);
- configuration of nanotube ends (e.g., open, capped);
- description of any branching;
- width/diameter of inner most wall/tube (average and range);
- carbon unit cell ring size and connectivity;
- alignment of nanotube along long axis (straight, bent, buckled);
- hexagonal array orientation used in the manufacture of the nanotube;
- particle size of catalyst used in the manufacture of the nanotube;
- molecular weight (average and range); and
- particle properties: shape, size (average and distribution), weight (average and distribution), count, surface area (average and distribution), surface to volume ratio, aggregation/agglomeration.

In April 2011 several members of the NanoSafety Consortium for Carbon, an association of businesses dealing in carbon nanomaterials formed in 2010, submitted a proposed testing agreement for substances including MWCNTs, double-walled CNTs, single-walled CNTs, and graphene nanoplatelets. Participants would conduct 90-day inhalation toxicity studies in rats, and submit interim progress reports to the EPA at 60-day intervals, and a final report within 120 days of the conclusion of testing. Under the proposed testing agreement, if the EPA promulgated a SNUR relating to the test substances, then the testing agreement would have the status of a TSCA section 5(b)(1)(A) test rule. The EPA had not responded at the time of writing.

7.1.3.4 Significant New Use Rules

Once a substance subject to a section 5(e) Order is listed on the TSCA Chemical Substance Inventory, any other company may manufacture the substance without being required to notify the EPA or comply with any other restrictions under TSCA section 5, unless the EPA promulgates a SNUR. Accordingly, when the

11. See for example, a redacted version of a 2008 Consent Order, at <www.nanolawreport.com/EPA%20Premanufacture%20Notice%20Number%20P-08-0177.pdf>.

EPA regulates a new substance under a risk-based section 5(e) Order, it develops a SNUR in parallel.[12] The SNUR defines a significant new use so as to require reporting to the EPA at least 90 days before any manufacturer or processor of the same chemical undertakes activities inconsistent with requirements of the section 5(e) Order.[13] The notice gives the EPA a basic set of information such as chemical identification, material characterization, physical/chemical properties, commercial uses, production volume, exposure and fate data, and toxicity data. On that basis the EPA can evaluate the intended uses of the nanoscale materials and where necessary take action to prohibit or limit activities that may present an unreasonable risk to human health or the environment.

To date the EPA has issued SNURs for less than 40 new chemical nanoscale materials. As a result of claims to the confidentiality of business information by submitters, the EPA is unable to identify the chemical substance as a nanoscale material in every new chemical SNUR it issues for nanoscale materials. However, in May 2011, the EPA promulgated a final SNUR (effective in June 2011) for the chemical substance identified generically as MWCNTs, which had previously been the subject of a PMN.[14] Under the final SNUR, persons intending to manufacture, import, or process MWCNT for a use that is designated as a significant new use by the final rule must notify the EPA at least 90 days before commencing that activity. Based on test data on analogous respirable, poorly soluble particulates and on other CNTs, the EPA identified concerns for lung effects, immunotoxicity and mutagenicity from exposure to MWCNTs.

The EPA is also developing a SNUR under section 5(a)(2) of TSCA to require notification of new nanoscale materials based on existing chemical substances – such as nanoscale silver, TiO_2 and ZnO_2.

7.1.3.5 Information Gathering Rule

Perhaps as a result of the lack of widespread support for the Nanoscale Materials Stewardship Program, the EPA is also developing a proposed rule to require the submission of additional information.[15] This rule would propose that persons who manufacture these nanoscale materials notify the EPA of production volume, methods of manufacture and processing, exposure and release information, and available health and safety data.[16] The data required could cover:

- The common or trade name, the chemical identity, and the molecular structure of each chemical substance or mixture.
- The categories or proposed categories of use of each such substance or mixture.

---

12. Found in 40 CFR Part 721.
13. Section 5(a)(1)(B) of TSCA.
14. P-08-199.
15. Under TSCA s. 8(a).
16. Although manufacturers can apply for exemption if they manufacture less than 500 kg, or manufacture only for research and development purposes.

- The total amount of each such substance and mixture manufactured or processed, reasonable estimates of the total amount to be manufactured or processed, the amount manufactured or processed for each of its categories of use, and reasonable estimates of the amount to be manufactured or processed for each of its categories of use or proposed categories of use.
- A description of the by-products resulting from the manufacture, processing, use, or disposal of each such substance or mixture.
- All existing data concerning the environmental and health effects of such substance or mixture.
- The number of individuals exposed, and reasonable estimates of the number who will be exposed, to such substance or mixture in their places of employment and the duration of such exposure.
- The manner or method of its disposal, and in any subsequent report on such substance or mixture, any change in such manner or method.

The rule had not been adopted at the time of writing.

7.1.3.6 Test Rule

The EPA is also proposing to introduce a rule to require testing for certain nanoscale materials – CNTs and nanosized clays and alumina – that are already in commerce.[17] The main objective is to obtain information about classes of nanoscale materials which are not already being tested by other US and international organizations, possibly including nanoclays and nanoscale aluminium. As well as assessing potential health and environmental risks, the results could also help to establish a correlation between the chemical/physical properties and the effects of the nanoscale materials.

The CNTs to be tested will be those with comparatively high production and high exposure. The tests will use OECD procedures, and may include a two-year bioassay, chronic exposure, and environmental fate testing. The latter will include leaching from landfills, weathering, incineration, and photolysis, since the EPA is concerned about the potential for CNTs that are agglomerated or entrained in the polymer matrix to be freed at the end-of-life and be released into the environment.

7.1.4 Environment

7.1.4.1 Federal Laws and Regulations

The EPA also enforces and drafts implementing regulations for most environmental protection laws of the United States, including the Clean Air Act, Clean Water Act, Safe Drinking Water Act, Resource Conservation and Recovery Act,

17. Under TSCA s. 4.

Comprehensive Response, Compensation and Liability Act (Superfund) and Federal Insecticide, Fungicide and Rodenticide Act (FIFRA).

Although none contain specific provisions on nanotechnology, each may affect the liability of producers and users of nanotechnologies within their scope. For instance, the pollutants regulated under the Clean Air Act include fine particles below 2.5 μm in size, which arguably includes nanoparticles even though they might more appropriately be categorized as ultra-fine particles, a category which is not separately listed.

The most significant regulatory development relating to nanotechnology in this context is the proposal of a new policy in relation to the implementation of FIFRA – under which pesticides must be registered before they can be sold in the United States[18] – which will address the use of nanoparticles in pesticides.[19] The EPA believes that information concerning pesticide ingredients which applicants and registrants have not previously been required to provide, is relevant to its statutory obligation to determine whether the registration of a pesticide may cause unreasonable adverse effects on the environment.

Accordingly, the EPA has solicited comment on two possible approaches to collecting information about what nanoscale materials are in pesticide products.

For the purposes of this proposed policy, the EPA has for the first time had to adopt a definition of 'nanoscale material'. In doing so, it reviewed the elements common to existing definitions, namely:

(1) the material's particle size measures typically between approximately 1 and 100 nm in at least one dimension;
(2) the material exhibits unique or novel properties compared to larger particles of the same material; and
(3) rather than occurring naturally, the material has been manufactured or engineered at the nanoscale to take advantage of these unique properties.

It concluded, however, that such a definition does not readily work in a regulatory context because of the high degree of subjectivity involved with interpreting such phrases as 'unique or novel properties' or 'manufactured or engineered to take advantage of these properties'. Moreover, the contribution of these subjective elements to risk has not been established. The EPA's proposed definition is therefore based on purely objective criteria, namely size. A material is a nanoscale material if it is:

> an active or inert ingredient of a pesticide and any component parts thereof intentionally produced to have at least one dimension that measures between approximately 1 and 100 nm regardless of the aggregation or agglomeration state of the final material.

In determining whether an ingredient meets this description, the EPA intends to review particle size data and also the manufacturing process to determine whether

18. Section 3(a) FIFRA.
19. Federal Register, 76, no. 117 (Friday, 17 Jun. 2011), 35383–35395.

it employs processes specifically to create or enhance the proportion of nanoscale materials in the product, as compared with other processes used to produce similar products. Although the definition would cover any chemical substance, the EPA also says that it expects these ingredients to be metal-based materials such as nano-silver and carbon-based materials such as CNTs. It specifically does not intend the definition to cover biological materials (e.g., DNA, RNA, proteins) or materials in their natural state (e.g., clays). Where it is unclear whether or not a particular material is within the definition or not, the EPA will review information as to the composition of the pesticide product and provide a view.

In order to grant a registration, the EPA must find that a product's composition warrants the proposed claims for it; that the product's labelling and other material required to be submitted comply with FIFRA; that the product will perform its intended function without causing unreasonable adverse effects on the environment; and that, when used in accordance with widespread and commonly recognized practice, the product will not cause unreasonable adverse effects on the environment.[20] 'Unreasonable adverse effects on the environment' include any unreasonable risk to man or the environment, taking into account the economic, social, and environmental costs and benefits of the use of any pesticide. What is reasonable therefore includes a cost-benefit analysis.

The two options which the EPA is considering for obtaining additional information specific to nanoscale materials are the use of data call-in notices[21] in respect of existing registrations or the obligation, under FIFRA section 6(a)(2), for registrants to inform the Agency of relevant information relating to their products, even though it was not specifically requested. The latter is the preferred option. The EPA has historically interpreted this to include not only information relating directly to adverse effects caused by pesticides, but also to other types of information and studies that the EPA would typically use in assessing whether a pesticide meets the statutory standard for registration.

It is not considering changing the data requirements generally for the registration of pesticides containing nanoscale materials across all the application and use scenarios as it does not consider that it has the knowledge base necessary to do so.

What is proposed, therefore, is that for purposes of registration under FIFRA section 3(c)(5) or 3(c)(7), the EPA would initially classify any application for registration of a pesticide product containing an active or inert ingredient that is a nanoscale material as an application for a 'new' active or inert ingredient, even when another registered pesticide product contains a non-nanoscale form of the ingredient or a nanoscale form of the ingredient with different size dimensions or other properties. This initial presumption could be rebutted on a case-by-case basis, for example through the submission of bridging data or other information demonstrating to the EPA's satisfaction that the nanoscale material's properties, which are relevant to assessing the potential risks to human health and the environment,

20. FIFRA s. 3(c)(5).
21. FIFRA s. 3(c)(2)(B).

are substantially similar to the properties of the already-registered form of the material, or that the nanoscale material differs only in ways that do not significantly increase the risk of unreasonable adverse effects on the environment. If so, then the application would be processed as a 'me-too' application within the abbreviated timeframes for such applications. If not, however, then the EPA would process the application as new active ingredients or new inert ingredients and would be likely to require the applicant to provide the types of data typically required for an assessment of the potential hazards and exposure to a new active or inert ingredient. Further, if a registrant wanted to change the composition of its product to include a nanoscale version of a material that the EPA had previously approved in non-nanoscale form, the registrant would need to notify the EPA and obtain approval before making the change.

Even once this document is finalized, the policy will have the status only of guidance to the EPA personnel and decision-makers and to pesticide applicants. Accordingly, the EPA would remain free to depart from the policy if and when circumstances warrant. Likewise, pesticide applicants would be able to assert that the proposed policy was not applicable to a specific pesticide or situation.

The comment period for the proposed policy expired on 17 August 2011.

7.1.4.2 State Laws

While the EPA sets national standards for implementing these laws, these standards are minimum standards. Individual US states are able to impose stricter standards within their own territories. A review of such laws is beyond the scope of this book.

7.1.5 Occupational Health and Safety

The Occupational Safety and Health Administration of the Department of Labor (OSHA) administers the Occupational Safety and Health Act[22] which aims to ensure safe and healthful working conditions for working men and women by setting and enforcing standards and by providing training, outreach, education and assistance.

Section 5(a) of the Occupational Safety and Health Act imposes a general duty on employers to provide employees with employment and a place of employment which are free from recognized hazards that are causing or are likely to cause death or serious physical harm. Employers must also comply with occupational safety and health standards promulgated under the Act. No nanotechnology-specific standards have yet been introduced, although various existing standards relating to issues such as personal protective equipment, general requirements, eye and face protection, respiratory protection, hazard communication and occupational exposure to hazardous chemicals in laboratories may be relevant.

22. 29 U.S.C. 654.

NIOSH provides detailed guidance on a number of workplace-related nanotechnology issues, most recently an Intelligence Bulletin on occupational exposure to titanium dioxide at both bulk and nanoscale[23]giving quantitative risk assessment using dose-response information from the rat and human lung dosimetry modelling and recommended occupational exposure limits for fine and ultra-fine (including engineered nanoscale) TiO_2, and exposure monitoring techniques, exposure control strategies, and research needs. A 2009 publication reviewed current knowledge of nanoparticle toxicity, process emissions and exposure assessment, engineering controls, and personal protective equipment.[24] This identifies a list of workplace tasks which increase the risk of exposure, and suggests approaches to manage exposure using techniques similar to those used in reducing exposure to aerosols in general, pending the development of sufficient information to monitor workplace exposures. NIOSH has also published interim guidance for medical screening and hazard surveillance for workers potentially exposed to engineered nanoparticles at work.[25]

These publications may affect the standard of care employers are required to achieve in order to meet the general duty required under section 5 of the Act.

7.2 CANADA

7.2.1 Overview

In 2008 the Expert Panel on Nanotechnology of the Council of Canadian Academies reported[26] that Canada's existing regulatory approaches and risk management strategies are appropriate for the challenges presented by nanomaterials, although recommending that more investment be made in strategic risk assessment research. The Federal Government's *Mobilizing Science and Technology to Canada's Advantage* strategy document of the same year committed the government to develop a plan to ensure that nanotechnology, along with biotechnology and information and communications technology, are regulated in a timely manner that draws on best practices.

The Canadian Federal Departments responsible for conducting risk assessments of nanomaterials concur that more information and experience is necessary, before it will be possible to determine whether or not new legislation or

---

23. *Current Intelligence Bulletin 63: Occupational Exposure to titanium diuoxide*, DHHS (NIOSH) Publication 2011-160.
24. *Approaches to safe nanotechnology: managing the health and safety concerns associated with engineered nanomaterials*. DHHS (NIOSH) Publication 2009-125.
25. <www.cdc.gov/niosh/review/public/115/>.
26. *Small is different: a science perspective on the regulatory challenges of the nanoscale*, prepared for the Government of Canada in response to a request from the Minister of Health via the Minister of Industry available at <www.scienceadvice.ca/uploads/eng/assessments%20and%20publications%20and%20news%20releases/nano/(2008_07_10)_report_on_nanotechnology.pdf>.

amendments to existing legislation are necessary. To that end, the Federal Government is funding a three-year study (initiated in 2008) to expand knowledge of some currently manufactured nanomaterials and to determine methods to monitor exposure to them from manufactured nanomaterials in chemical substances and Food and Drugs Act substances. Analytical methods to characterize the nanomaterials, and methods to monitor exposure to them, are being analysed. Preliminary testing is also being performed to examine any potential health or environmental effects. This project is intended to serve as a gateway for more advanced research to characterize the loadings, fate, and exposure pathways for contamination by nanomaterials.

In the meanwhile, nanomaterials are regulated in Canada under existing legislation, administered by Health Canada and Environment Canada.

7.2.2 Food and Drugs

7.2.2.1 Applicable Legislation

Health Canada regulates a wide range of products and substances that are or could be affected by nanotechnology. For instance, under section 4(1) of the Food and Drugs Act no person shall sell an article of food that has in or on it any poisonous or harmful substance. Although nanomaterials are not mentioned, clearly any nanomaterial which was harmful to health would be prohibited by this provision. Ignorance of harmful effects is not generally considered a defence where consumer protection legislation is concerned. As to food additives, only those additives listed in the Tables of Division 16 of the Food and Drug Regulations are permitted,[27] and may be used only for their permitted uses. Individual food additives may not be used at levels greater than their respective inscribed maximum levels of use. If an additive is not listed, then the manufacturer should request a Letter of Opinion from Health Canada's Bureau of Chemical Safety as to its regulatory status. The list of permitted additives includes both carbon black and titanium dioxide without stipulating the scale,[28] consistent with Environment Canada's approach to materials on the Domestic Substances List (discussed at section 7.2.3 below).

Likewise, Regulation 24 of the Cosmetics Regulations[29] requires the label of a cosmetic that presents an avoidable hazard to include directions for safe use. An 'avoidable hazard' is defined as a threat of injury to the health of the user of a cosmetic that can be:

(a) predicted from the cosmetic's composition, the toxicology of its ingredients and the site of its application;
(b) reasonably anticipated during normal use; and

27. Regulation B.16.100.
28. Table III.
29. C.R.C., c. 869, under the Food and Drugs Act.

(c) eliminated by specified limitations on the usage of the cosmetic.

Once again, although neither nanotechnology nor nanomaterials are specifically mentioned, this provision would require any manufacturer who was using in a cosmetic preparation a nanomaterial with known or predictable toxicity through external use to include specific directions for safe use on the label (along with the complete list of ingredients required by Regulation 21.2).

Health Canada also has responsibility under Environmental Protection Act of 1999[30] (CEPA) to assess and manage risks to human health from the use of new substances as well as the environmental risks for substances under the Food and Drugs Act.

7.2.2.2 Definition of Nanomaterials

In view of the potential impact on its various areas of responsibilities of nanomaterials, in 2010 Health Canada adopted an *Interim Policy Statement on Health Canada's Working Definition for Nanomaterials*,[31] with effect from the date of posting. It is intended to be updated as necessary. The objective is to establish a working means for identifying materials, achieve consistency between departments, support clarity of communications about nanomaterials (including requests for information) and assist Health Canada to collect information and establish internal inventories relating to products, materials, ingredients and devices that are, contain or use nanomaterials.

The definition relates only to manufactured products and materials, so naturally occurring nanomaterials are excluded, as are nanoparticles produced as non-intentional by-products of a process. Under the definition, any manufactured product, material, substance, ingredient, device, system or structure is considered to be nanomaterial if it is:

- at or within the nanoscale in at least one spatial dimension, or;
- smaller or larger than the nanoscale in all spatial dimensions and exhibits one or more nanoscale phenomena.

For the purposes of this definition:

- 'nanoscale' means 1 to 100 nm, inclusive;
- 'nanoscale phenomena' means properties of the product, material, substance, ingredient, device, system or structure which are attributable to its size and distinguishable from the chemical or physical properties of individual atoms, individual molecules and bulk material; and
- 'manufactured' includes engineering processes and control of matter and processes at the nanoscale.

---

30. 1999, c. 33.
31. Available at <www.hc-sc.gc.ca/sr-sr/alt_formats/pdf/consult/_2010/nanomater/draft-ebauche-eng.pdf>.

Note that a material does not actually have to have *any* dimension in the nanoscale, provided it 'exhibits nanoscale phenomena', of which there is no definition (and may be little international scientific consensus). The breadth of the working definition is intentional in view of the range of nanomaterial-related regulatory responsibilities at Health Canada. The Policy Statement indicates that it will be applied more specifically in each regulatory programme area. This introduces an unfortunate uncertainty for commercial undertakings and means that consultation with the relevant regulatory or supervisory bureau is still needed, in order to avoid including more substances or products as nanomaterials for notification or reporting purposes than may in practice be required. This precautionary communication is itself an objective of the broad definition, since Health Canada aims to increase the information available to it for policy and regulatory decisions.

Information which may be required includes:

- the intended use of the nanomaterial, including the end product;
- characterization of the nanomaterial including manufacturing methods, identity and purity;
- physico-chemical properties, toxicological, eco-toxicological, metabolism and environmental fate data (both generic and specific); and
- any risk assessment and risk management strategies which have been considered or implemented.

The Policy Statement indicates that specific guidance documents will be developed over time, and that it is intended to develop a consistent set of approaches across the various programme and regulatory areas, but that the precautionary approach will be applied.

7.2.3 Chemicals

Environment Canada is responsible for administering, under CEPA a legal regime for the assessment and management of new and existing substances.[32]

CEPA serves as a safety net by requiring environmental and health assessments for substances and products that are not regulated by other legislation. It requires importers or manufacturers of new substances to provide specific information to government so that they can be evaluated for potential effects on human health and the environment.[33] The definition of substance is:

> any distinguishable kind of organic or inorganic matter, whether animate or inanimate, and includes
>
> (a) any matter that is capable of being dispersed in the environment or of being transformed in the environment into matter that is capable of being

32. Part 5 of the Act.
33. Section 81(1) CEPA.

so dispersed or that is capable of causing such transformations in the environment,

(b) any element or free radical,

(c) any combination of elements of a particular molecular identity that occurs in nature or as a result of a chemical reaction, and

(d) complex combinations of different molecules that originate in nature or are the result of chemical reactions but that could not practicably be formed by simply combining individual constituents.[34]

This is sufficiently broad to include nanoforms of known substances as well as new substances having significant nanoscale properties. The only exclusions are nanomaterials which are regulated under other acts such as the Pest Control Products Act, the Fertilizers Act and the Feeds Act. Where a nanoscale form of a substance which in bulk form is already on the Domestic Substances List (DSL, of chemicals already in use in Canada, as periodically updated) has unique structures or molecular arrangements then it is considered a 'new' substance. Otherwise, it will not be. For example, titanium dioxide is listed on the DSL and since its nanoscale form does not have unique structures or molecular arrangements, it is not subject to the New Substances Notification Regulations. Nor are incidentally produced or naturally occurring nanomaterials subject to notification.[35]

The details of the notification process and information requirements are set out in the New Substances Notification Regulations (Chemicals and Polymers) and accompanying comprehensive *Guidelines for the Notification and testing of New Substances: Chemicals and Polymers*.[36] For substances neither on the DSL nor on the US Toxic Substances Inventory List,[37] the amount triggering a notification requirement can be as little as 100 kg. In 2010, Environment Canada and Health Canada undertook a mandatory information gathering survey, under the authority of section 71 of CEPA, to gather use pattern information, including volumes and sectors of use, and any relevant toxicological data already available for nanomaterials which had been in commercial use in Canada during 2008. The results of the survey had not been published at the time of writing.

Although not required, Environment Canada and Health Canada recommend that notifiers request a pre-notification consultation during the planning or preparation of a notification relating to nanomaterials. For example, a notifier can request a consultation to assist with determining whether the substance is

34. Section 3 CEPA.

35. New Substances Program Advisory Note 2007-06 Requirements for nanomaterials under the *New Substances Notification Regulations (Chemicals and Polymers)* at <www.ec.gc.ca/subsnouvelles-newsubs/default.asp?lang=En&n=3C32F773-1>.

36. Respectively, SOR/2005-247) available online at <http://laws-lois.justice.gc.ca/eng/regulations/SOR-2005-247/index.html> and Version 2005- EPS M-688 available online at <www.ec.gc.ca/subsnouvelles-newsubs/default.asp?lang=En&xml=FA59945E-8DD5-D745-8A55-EA6667737FE5>.

37. Which is used as the basis for the Non-Domestic Substances List, of substances which while not in use in Canada are known to be in use internationally.

notifiable, as well as to clarify notification procedures or information requirements, and to determine the acceptability of waiver requests and/or test protocols.

In 2010 a Private Member's Bill[38] was introduced to amend CEPA to require risk assessment procedures prior to nanomaterial or nanoproduct release into the marketplace, the environment or to Canadians. Although as a Private Member's Bill it is unlikely to be passed, its sponsor reports that support is growing for some of the concepts incorporated in the Bill, including the importance of a public inventory of nanotechnology and nanomaterials in Canada.[39]

7.2.4 Environment

Environment Canada's responsibility in respect of environmental laws other than chemicals management covers environmental and human health impacts of products of biotechnology, marine pollution, disposal at sea, vehicle, engine and equipment emissions, fuels, defined hazardous wastes, environmental emergencies and international affairs, with no provisions specific to nanotechnology in relation to any of these.

Specific provincial legislation regulates air quality, water, waste and so on for each province. A review of such laws is beyond the scope of this book.

7.2.5 Occupational Health and Safety

Section 124 of the Canada Labour Code[40] imposes on employers a general duty to ensure that the health and safety at work of every employee is protected. More specific duties are set out in sections 125 and 125.1. These include duties to implement and monitor a programme for hazard prevention, including educating employees in health and safety matters related to those hazards. Further, where hazardous substances are present in the work place the employer must ensure that their concentrations are controlled in accordance with prescribed standards; and they are identified, stored and handled in the manner prescribed. To date there are no standards specific to nanotechnology, but the Canadian Standards Association has formed a Technical Committee on Nanotechnologies – Occupational Health and Safety. Initial activities include working to adopt the published international ISO Technical Report, ISO/TR 12885:2008 on Health & Safety Practices in Occupational Settings relevant to Nanotechnologies, as well as to produce a national standard to provide guidance for workers using nanomaterials in the workplace.

38. Bill C-494.
39. <www.frogheart.ca/?p=2841>, *Canada's Bill C-494, nanotechnology safety in Canada: an update,* reporting an update from Peter Julian MP, February 2011.
40. R.S.C., 1985, c. L-2.

7.3 AUSTRALIA

7.3.1 Food and Drugs

The Australian food regulatory scheme was created in the absence of any consideration of nanotechnology, and does not mention it, but may nevertheless be applied to nanofoods and related products. All food supplied in Australia and New Zealand must comply with the Australia New Zealand Food Standards Code (the ANZF Code) and be safe for human consumption. Any new food substances that are manufactured using nanotechnologies and may present safety concerns will have to undergo a comprehensive scientific safety assessment by Food Standards Australia New Zealand (FSANZ) under the appropriate Standard before they can be legally supplied in Australia. Anyone applying for a food standard must declare any nanosized particle in their application, but as at May 2011 FSANZ had not yet received any applications to approve any new type of nanoparticles in foods.[41] There is no certainty as to whether a nanoform of an existing food substance is to be treated as a new food substance or not. Nor does the Code address the question of naturally nanoform substances in foods.

An application for pre-market approval must be submitted to FSANZ in accordance with the FSANZ Application Handbook. This was reviewed in 2008 to address concerns that it may not apply risk assessment methodologies appropriate for nanoscale materials,[42] and amended in December 2008 to incorporate the essential information required to make an application to vary the ANZF Code (to add a new substance to the permitted substances list). Information on particle size, size distribution and morphology is needed where substances are particulate in nature and will remain so in the final food. The amendment applies to food additives, processing aids, novel foods, contaminants and nutritive substances. Applications to approve a food substance must also include the raw data and studies necessary to conduct a comprehensive risk assessment and establish the safety of that specific food substance. If there is insufficient data to establish the safety of the food substance, FSANZ will not grant approval.

Through the standards in the Code, FSANZ has the capacity to prohibit or grant different permissions for different physical forms of food substances, should the safety assessment determine there is a case to do so. The key issue is whether there is sufficient evidence to establish the safety of the exact physico-chemical entity.

The Code also sets out the requirements for labelling foods in Australia and New Zealand. An independent review of Australia's food labelling requirements was carried out in 2010, and reported in January 2011. The report[43] recommends

---

41. <www.foodstandards.gov.au/consumerinformation/newfoodsandtechnolog4358.cfm>.
42. See for instance, Ludlow, *The Readiness of Australian Food Regulations for the Use of Nanotechnology in Food and Food Packaging*. UMonashLRS 14 (2009).
43. *Labelling Logic – The Final Report of the review of Food Labelling Law and Policy*, available at <www.foodlabellingreview.gov.au/internet/foodlabelling/publishing.nsf/Content/labelling-logic>.

that as a general principle all foods or ingredients that have been processed by new technologies (i.e., all technologies that trigger pre-market food safety assessments) be required to be labelled for 30 years from the time of their introduction into the human food chain; the application of this principle to be based on scientific evidence of direct impact on, or modification of, the food/ingredient to be consumed. At the expiry of that period the mandatory labelling should be reviewed. This clearly would cover nanotechnology alongside other technologies such as genetic modification of food or feeds.[44] Further, for the credibility of the regulator the report recommended that a standard be established for regulating the presence of nanotechnology in the food production chain, despite the risk that such labelling may create an unjustified public perception that all nanoparticles/nanomaterials are hazardous.[45] The report is now under consideration by the government and may lead to further amendments to the Code.

Drugs are regulated under the Therapeutic Goods Act 1989,[46] and by section 25 must be evaluated before marketing as to whether the quality, safety and efficacy of the goods for the purposes for which they are to be used have been satisfactorily established, which would require the showing of non-toxicity for nanoparticles or delivery systems. There are no nano-specific standards or regulations in effect.

7.3.2 Chemicals

7.3.2.1 Industrial Chemicals

Chemicals regulation in Australia is integrated under the National Industrial Chemicals Notification and Assessment Scheme (NICNAS), which is responsible for assessing all chemicals new to Australia before they can be placed on the market, and also those chemicals already used, in response when concerns are raised about their safety on health and environmental grounds.[47]

Following consultation, NICNAS has enacted a new regulatory process specifically addressing nanotechnology, making Australia the first developed country to do so. The new process took effect on 1 January 2011, and applies to any new material/chemical that falls under the working definition of an industrial nanomaterial:

> industrial materials intentionally produced, manufactured or engineered to have unique properties or specific composition at the nanoscale, that is a size range typically between 1 nm and 100 nm, and is either a nano-object

44. Recommendation 28.
45. Recommendation 35.
46. Act No. 21 of 1990.
47. The relevant legislation is the Industrial Chemicals Notification and Assessment Act 1989 (ICNA).

(i.e., that is confined in one, two, or three dimensions at the nanoscale) or is nanostructured (i.e., having an internal or surface structure at the nanoscale).

Any substances meeting the definition that are new chemicals, that is nanoforms of chemicals not listed on the Australian Inventory of Chemical Substances, will require a NICNAS permit or certificate (as do all new chemicals) but will not be eligible for those certificate exemption categories where human and/or environmental exposure can reasonably be anticipated, namely the low volume cosmetic and non-cosmetic exemptions[48] and the low concentration (<1%) non-hazardous cosmetic exemption.[49] Any applicant claiming one of these exemptions must therefore now also state that their product is not a nanomaterial.

The transhipment exemption remain unchanged as does the research and development exemption,[50] with amendments to the annual reporting requirements. All nanomaterials introduced in volumes over 100 g will have to be identified as nanomaterials and their full chemical name provided.

Additional information will also be required in respect of permit applications, including a declaration by the notifier on the permit application forms as to whether the chemical is a nanomaterial or not. More specific information (such as particle size, shape and other specific information on properties) may be required where NICNAS considers it necessary for the purpose of assessing the chemical. For example, where the nanomaterial is introduced as a solid/powder or as a dispersion and is insoluble or known to be biopersistent, then as a minimum requirement the primary particle size distribution of the nanomaterial will be required. Further additional data may be required such as surface area, impurity profile and surface properties (such as charge and coatings), on a case-by-case basis.[51]

The explanatory notes[52] specify that the definition is not intended to cover nanoscale materials which are produced accidentally as part of the production of a bulk-scale form of a substance. In such a case, the proportion of total number of particles at the nanoscale is likely to be relatively small. However, where a particle size distribution shows 10% or more of a substance is at the nanoscale, NICNAS will consider this substance to be a nanomaterial for risk assessment purposes.

The definition is expressly a working definition which may (and probably will) be amended following a reasonable period in operation, as scientific understanding of nanomaterials and the behaviour in the body and the environment increases.

Until then, the controversial issue of precisely what amounts to the nanoscale has been left open by the use of the qualification 'typically', thereby avoiding the

48. Section 21(4) ICNA.
49. Section 21(6c) ICNA.
50. Section 21(6a)).
51. The details of the administrative regulations for this purpose are available at <www.nicnas.gov.au/Publications/Chemical_Gazette/pdf/2010oct_whole.pdf#page=14>.
52. *New Chemicals Requirements for Notification of Industrial Nanomaterials*, in the *NICNAS Notifiers Handbook*.

debates as to whether a substance at 101 nm is still nanoscale. This is a pragmatic approach since the scientific debate over nanomaterials and their characterization is still at an early stage, although leaving a measure of uncertainty for manufacturers and importers needing to decide whether their materials require a permit or not.

The unique properties referred to are intended to be chemical and/or physical properties that are different because of the substance's nanoscale features as compared to the same material without nanoscale features, and that result in unique phenomena (e.g., increased strength, chemical reactivity or conductivity) that enable novel applications.

The definition is also intended to cover aggregates and agglomerates as being nanostructured substances. This addresses one of the major concerns of environmental and consumer groups, that aggregates and agglomerates which may overall have dimensions of around 300 nm could as a result be excluded from regulatory scrutiny whilst still possessing the special characteristics of nanomaterials which lead to toxic or eco-toxic properties.

7.3.2.2 Pesticides and veterinary medicines

Separately, the Australian Pesticide and Veterinary Medicines Agency is assessing whether it needs to modify its data requirements and the risk assessment framework to address agricultural and veterinary chemicals and chemical products containing engineered nanomaterials. Registration of such products is done individually, and any change to the composition or form of a registered chemical or chemical product creates a new chemical or chemical product requiring new assessment.

7.3.3 Environment

7.3.3.1 Environment Protection and Biodiversity Conservation Act

Environmental issues are primarily dealt with in the Environment Protection and Biodiversity Conservation Act 1999 (the EPBC Act). This was subject to an independent review in 2008, and the report,[53] issued in 2009, included a number of observations on nanotechnology. These were predictably precautionary. The report noted that given the early stage in nanotechnology development, it would be prudent to bear in mind the impact such emerging technologies may have on the environment and how they may be addressed under the EPBC Act. Processes associated with the manufacture, use or disposal of nanomaterials could be considered as actions that may have a significant impact on matters protected

53. *Report of the Independent Review of the* Environment Protection and Biodiversity Conservation Act 1999 Ch. 6, *148,* 30 Oct. 2009, Dr Allan Hawke.

under the EPBC Act. It is important that the EPBC Act be able to assess the risk posed by these materials on matters protected under it.

The report's author recognized that GMO and nanotechnology assessments under the EPBC Act pose some difficulties – for example, it may be difficult to identify who is taking an action where multiple parties may have taken a collective action such as releasing a product that affects a protected matter some distance from the area of release. The most effective method would be to identify the technology provider as the party taking the action. However, as they are not physically taking the action, this approach may not work under the current provisions of the EPBC Act.

Consequently, the report recommends that the EPBC Act needs to be able to identify as persons taking an action, the developers/providers of products derived from emerging technologies that may disperse or disseminate and impact on a protected matter. It does not make any more specific recommendation for amendment of the Act or its administration.

7.3.3.2 Product Stewardship Bill 2011

The Product Stewardship Bill 2011, introduced into Parliament in March 2011, is intended to set up a national 'cradle-to-grave' regime to manage the environmental, health and safety impacts of products, and in particular those impacts associated with the disposal of products. The framework includes voluntary, co-regulatory and mandatory product stewardship, the object being to provide first for voluntary and co-regulatory schemes rather than imposing a mandatory scheme from the outset. The Bill will provide the basis for obligations to be imposed on manufacturers, importers, distributors and others to take action to reduce or eliminate hazardous substances in products and waste from products; and ensure that products and waste from products is treated, disposed of, recovered, recycled and reused in a safe, scientific and environmentally sound way. However, as the Bill is designed to establish a legislative framework for product stewardship schemes, it is silent on the products that may be covered at some point in the future. Different industry sectors will decide which products they seek accreditation for under the voluntary arrangements, while it will be for the environment minister to decide which products to regulate under the co-regulatory or mandatory arrangements.

If passed, the Bill would use product stewardship criteria to determine whether its provisions should apply to a particular class of products. Before regulations are made for either a co-regulatory or mandatory scheme, the minister must first be satisfied that the regulations will further the objects of the Bill and that the product stewardship criteria are met.

The product stewardship criteria, outlined in clause 5, are satisfied if two or more of the following apply to a class of product:

(a) the products are in a national market;
(b) the products contain hazardous substances;
(c) there is the potential for resource recovery benefits;

(d) reusing or recycling of the products involves a significant cost to the Commonwealth, state, territory or local governments;
(e) consumers' willingness to pay; and
(f) taking action will offer business opportunities that would make a contribution to the economy.

These criteria could conceivably require all household items sold in Australia to be covered by a product stewardship scheme, creating uncertainty for businesses until the government's approach to mandatory regulation becomes clearer.

The Senate Environment and Communications Legislation Committee released its Report on the Bill[54] in May 2011, recommending that the Bill be passed in the Senate's winter 2011 session.

7.3.4 Occupational Health and Safety

Nanotechnologies and engineered nanomaterials are considered already to be regulated under Australian work health and safety regulations for hazardous chemicals generally. Nonetheless, Safe Work Australia (SWA) is advocating a precautionary approach be taken to controlling exposures to engineered nanomaterials and is in the process of revising the Code of Practice for SDS, proposing the addition of a number of non-mandatory parameters specifically relevant to engineered nanomaterials in physico-chemical sections. Many engineered nanomaterials are not currently classified as hazardous so it is not mandatory to prepare an SDS or include information on labels. However, SWA recommends that for engineered or manufactured nanomaterials an SDS/label be provided unless there is evidence they are not hazardous.

The revised Code of Practice will include additional non-mandatory parameters under physical and chemical properties, namely:

- Particle size (average and range), size distribution and surface area
- Shape and aspect ratio
- Crystallinity
- Dustiness
- Degree of aggregation or agglomeration, and dispersibility
- Biodurability or biopersistence
- Surface coating or chemistry (if different to rest of particle).

SWA's objective is to have model legislation – a Work Health and Safety (WHS) Act and Regulations covering Work Health and Safety (Hazardous Chemicals) – adopted in all states to harmonize legislation by the end of December 2011.

---

54. <www.aph.gov.au/senate/committee/ec_ctte/product_stewardship/report/report.pdf>.

7.4 JAPAN

7.4.1 Food and drugs

The Japanese Ministry of Health, Labour and Welfare (MHLW) is responsible for food standards, pharmaceuticals and cosmetics, under various statutes.

The Pharmaceutical Affairs Law of Japan regulates pharmaceuticals, medical devices, quasi-drugs, and cosmetics. In principle, it stipulates the submission of applications for the approval of manufacture and sales for all pharmaceuticals, medical devices, and quasi-drugs before commercialization, and the efficacy, safety and other aspects are confirmed for each product in advance. For all products, including cosmetics, any adverse events generated after product releases must be reported according to established rules to permit the implementation of appropriate safety measures. There are no explicit references to nanotechnology in the law, nor are there regulations specific to nanotechnology issues; rather, the issues are handled in the course of evaluation of each product on a case-by-case basis. The definition of pharmaceuticals in the Law includes all items, other than machines, instruments, or the like, (excluding quasi-drugs), used for the diagnosis, treatment, or prevention of human or animal diseases.[55] Being a functional definition, this clearly includes all forms of nanotechnology where these are used for those purposes. MHLW has supported various research programmes related to nanomedicine since 2002, and in 2008 a new section has been established at the National Institute of Health Sciences on evaluating nanomedicine.

The definition of quasi-drugs is likewise functional – including products other than machines, instruments, or other such products, having mild effects on the human body, and equivalent items designated by the Minister of Health, Labour and Welfare (such as medicated cosmetics, hair dyes, vitamin preparations) used to

(i) prevent nausea, discomfort, foul breath, or noticeable body odour;
(ii) prevent heat rash, sores, etc.;
(iii) prevent loss of hair, promote hair growth, or to remove hair;
(iv) to kill or repel rats, flies, mosquitoes, fleas, etc. and thereby to protect humans or animals.[56]

Any nanotechnology having mild effects and used for one of the listed functions, would be covered and require pre-market approval.

Cosmetics are regulated by Article 2–3 of the Pharmaceutical Affairs Law and defined as:

> products having mild effects on the human body (excluding medicines and quasi-drugs) and intended for cleansing, beautification, or enhancement of attractiveness and used to modify appearance, maintain the youthful appearance of skin, or to maintain healthy hair by means of application on the body.

55. Article 1 Pharmaceutical Affairs Law as amended.
56. Article 2 Pharmaceutical Affairs Law as amended.

Once again, nanotechnologies used for these purposes would be regulated.

Cosmetics ingredients are addressed by cosmetics standards. There is a negative list, of prohibited ingredients, and a positive list of usable ingredients. Addition of a new ingredient to the positive list requires a review and safety assessment by MHLW. All ingredients must be indicated on the packaging.[57]

The safety of foods in general is the responsibility of food-businesses which are required under the Food Sanitation Law to take measures to ensure that foods for human consumption are safe. In addition to disclosing the name of any food product which is suspected to be harmful to health, the MHLW can prohibit the sale of such food product in accordance with regulations specified in the Food Sanitation Law.

Food additives and packaging are also regulated under the Food Sanitation Law. This includes both substances remaining in the finished food products, such as preservatives, and substances not remaining in the finished products, such as infiltration-supporting agents. Food additives are permitted for use based on the completion of prior risk evaluations for each food additive.

In principle, the additives used in or on food must be clearly indicated on food products. Acceptable daily intake (ADI) levels are set for each food additive, based on scientific data, such as results of substance analysis and animal toxicity tests. Criteria for use are determined based on these ADI and other parameters.

There are also set specifications and criteria for containers and packaging, including the prohibition of packaging which contains or is covered with any harmful substance and involves a risk to human health. Containers and packaging that fail to comply with such specifications or criteria may not be sold. Local government offices are required to report to MHLW any report of damage to a consumer's health potentially attributable to food.

If any serious accidents occur due to consumer products, including household products, businesses must report to the Japanese Ministry of Economy, Trade and Industry (METI) in accordance with the Consumer Products Safety Act. Of these reports, those deemed to require response, including those indicating potential connections between chemical substances and the accidents, are reported to MHLW to publicize the issue and take other actions, as necessary.

7.4.2 Regulation of Chemicals

Japan's Act on the Evaluation of Chemical Substances and Regulation of their Manufacture, etc. (Chemical Substances Control Law or CSCL) was first enacted in 1973 in the wake of environmental pollution caused by PCBs (polychlorinated biphenyls). It sets up a system of pre-market assessment of industrial chemicals, administered by METI, which is intended to prevent pollution by chemical substances with similar properties to PCBs, including high persistency, high accumulation, and long-term toxicity. Newly manufactured or imported chemical

57. Article 61 of the Pharmaceutical Affairs Law.

substances with PCB-like properties are regulated for their manufacturing, import and use. In 1986, the law was amended to include the regulation of chemical substances that are not highly accumulative but that are persistently toxic (Type II Specified Chemical Substances and Designated Chemical Substances – current Type II Monitoring Chemical Substances). In 2003, the law was further amended introducing an assessment and regulation system to focus on the impact of chemical substances on flora and fauna, and the possibility of discharge into the environment.

In 2009, further amendments were introduced which require companies to report hazard and use information for chemicals manufactured in or imported into Japan. The objective was to move the Japanese chemical control system from a hazard-based to a risk-based approach. The reporting threshold is 1 tonne per year per company. The Japanese authorities will use the new information to prioritize certain chemicals for risk assessment and may request further exposure and hazard information from the companies that manufacture or import those substances. Chemicals found to be of highest concern will be virtually prohibited, with prior permission required to continue any specific uses. A revised system for collecting hazard and use data came into full effect on 1 April 2011.[58] Companies are required to report production or import annually with effect from 2011.

The application of this regime to nanomaterials is not entirely clear since the definition of 'chemical substance' is:

> 'a chemical compound obtained by causing chemical reactions to elements or compounds',[59]

subject to various exceptions covering radioactive substances and substances which are subject to other equally or more stringent regulations (Poisonous and Deleterious Substances Control Act) or more specific to the usage of the substance (Food Sanitation Act). On a strictly literal interpretation, this definition does not appear to include nanomaterials, either where they are single-element structures such as the various CNTs or even, arguably, where the nanoform is created by mechanical or electrical rather than chemical means, if the bulk form is already reported.

However, a nanomaterial which is also a new chemical substance would be covered. Some notifications of manufactured nanomaterials (e.g., fullerene derivatives) have reportedly been observed in the small quantities exemption notification scheme under the Law.

In November 2008, METI organized a study group (The Expert Meeting on Safety Measures for Nanomaterial Manufactures, etc.) to assess the safety measures voluntarily introduced by nanomaterials manufactures. The review concentrated on major nanomaterials: CNTs, carbon black, titanium dioxide, fullerenes,

58. Details are available at <www.meti.go.jp/policy/chemical_management/english/documents.html>.
59. CSCL, Ch. 1 Art. 2.

zinc oxide, and silicon dioxide. A study report published in March 2009[60] recommended enhanced but voluntary transmission of safety and toxicity data between manufacturers and by manufacturers to government for publication. There were no recommendations for regulatory or legislative changes.

7.4.3 Environment

Japan's environmental laws include restriction of industrial emissions, restriction of products, restriction of wastes, improvement of energy conservation, promotion of recycling, restriction of land utilization, arrangement of environmental pollution control programmes and provision for sanction. The 1993 Basic Environmental Law[61] places responsibility on corporations for taking necessary measures to prevent environmental pollution, such as the treatment of smoke and soot, polluted water and wastes resulting from their activities, and to properly conserve the natural environment, pursuant to the basic principles.[62] It also requires the government to establish environmental quality standards maintenance of which is desirable for the protection of human health and the conservation of the living environment.[63]

The Prime Minister's Office is authorized to establish maximum levels for discharges of into water of substances designated by the Cabinet Office as harmful as a result of their capacity to cause damage to the living environment.[64] Prefectural governors have power to order measures to prevent soil contamination by substances designated by the Cabinet Office as being likely to bring harmful effects on human health because of the existence of such substance in soil.[65] Similarly, the Cabinet Office can designate a substance as a 'designated particulate' for the purposes of the Air Pollution Control Law[66] if it is liable adversely to affect human health, in which case it is necessary to notify the prefectural governor of the nature of the facility, the method of operation and the method of disposal or control of scattering of the designated particulate, before installing a facility which may discharge such particles.[67] It does not appear, however, that any such designations have yet been made by the Cabinet Office in respect of nanoforms of substances.

In March 2009 the Japanese Ministry of the Environment issued voluntary guidelines aimed at reducing the risk of harm to the environment through the

60. Available at <www.meti.go.jp/policy/chemical_management/files/The%20Expert%20Meeting%20on%20Safety%20Measures%20for%20Nanomaterial%20Manufactures%20etc..pdf>.
61. Law No. 91 of 1993.
62. Article 8 Basic Environmental Law.
63. Article16 Basic Environmental Law.
64. Article 3, Water Pollution Control Law No. 138 of 1970 Latest Amendment by Law No. 75 of 1995.
65. Soil Contamination Countermeasures Act (Act No. 53 of 2002).
66. Law No. 97 of 1968 Latest Amendment by Law No. 32 of 1996.
67. Article 18-6 of the Air Pollution Control Law.

release of nanomaterials.[68] The guidelines recommend preventing the exposure of humans, plants and animals to manufactured nanomaterials by identifying the potential release route into the environment. Based on this, manufacturers are advised to build manufacturing equipment or facilities that do not release manufactured nanomaterials, design products that separate manufactured nanomaterials from other materials, and control manufactured nanomaterials to prevent their release. If there is a possibility of manufactured nanomaterials being released into the environment, a process to capture and remove them during the release route is required. If manufacturing equipment or facilities do not have measures in place to prevent the release of manufactured nanomaterials into the environment, the use of substitute materials should be considered, unless the manufacturers can fully assess the safety risk caused by the release of the manufactured nanomaterials.

7.4.4 Occupational Health and Safety

The Labour Standards Bureau of MHLW is responsible for preventing exposure to nanomaterials at workplaces, and promulgated a notice on Measures for Prevention of Exposure, etc. to Nanomaterials at Workplaces[69] in 2009. This addresses best practice for manufacturing or handling of nanomaterials and other materials containing nanomaterials (including manufacturing for research purposes, repairs and inspections of equipment or instruments in which nanomaterials are used), or disposal or recycling of products containing nanomaterials. It is advisory only: enterprises which have specific information which could be used to prevent exposure are expressly directed to take their own measures.

Enterprises are directed to gather detailed information about nanomaterials, etc. which are handled, such as data on electron micrographs, particle size, and specific surface area. Suppliers are required to transfer information to the customer to assist workers in preventing exposure, including the name and components of and precautions for handling such nanomaterials, by labelling the container or package, issuing a material SDS or by other ways.

7.5 CHINA

7.5.1 Overview

China was an early leader in the adoption of national nanotechnology-specific standards,[70] arising from a series of state-funded research programmes. It has

68. Available at <www.env.go.jp/chemi/nanomaterial/eibs-conf/guideline_0903_enab.pdf>.
69. Labour Standards Bureau of the Ministry of Health, Labour and Welfare, Notification No. 0331011, 31 Mar. 2009.
70. 12 were in effect by 2007; see country report for the OECD WPMN <www.oecd.org/dataoecd/23/52/39149145.pdf>.

been estimated that China's spending on nanotechnology research is second only to that of the United States. However, it has not to date enacted any national legislation or regulation specific to nanotechnology, other than its use in medical devices, and the lack of transparency of China's administration generally makes it difficult to establish how issues arising from the production and use of nanotechnology are being addressed within the existing legislative framework.[71]

The China Nanosafety Laboratory, established in 2006, has amongst other responsibilities methodological and metrological studies of nanoparticle detection; identification and quantification of nano-hazards to humans and the environment; establishing standard procedures for safety assessment of nanoproducts for nano-industries/enterprises including assessment methods, and identifying the toxicity classes of nanomaterials; and drafting regulatory frameworks for research and industrial activities on nanotechnology.

In the area of health and safety, China has a complex regime to manage and oversee the use and manufacture of chemical substances,[72] with multiple ministries including the Ministries of Health, Communication, Public Security and multiple administrative organs including the State Food and Drug Administration, State Administration of Standardization, State Administration of Work Safety all having some control over environment, health and safety aspects of chemicals. Where there may be regulatory gaps, as in determining whether regulations apply to nanotechnology and if so how, it is relatively harder to establish where responsibility lies for addressing this and there is a risk of inconsistent approaches being adopted. The difficulty is compounded by the enforcement regime, under which day-to-day enforcement of chemical production and food safety regulations fall directly onto local governments, which may inconsistently implement or interpret the national laws.

7.5.2 Food and Drugs

The only area in which any direct regulation has been introduced relating to nanotechnology is that of medical devices. In 2006 the State Food and Drug Administration revised its medical device regulations[73] to require medical devices made with nanometre biological materials and medical instruments made with nanometre metal silver material to be classified as Class III medical devices.

71. This section is based upon work carried out as a joint project between the London School of Economics and the Lee Kuan Yew School of Public Policy, National University of Singapore and in particular the paper *Mapping Emerging Nanotechnology Policies And Regulations: The Case of the People's Republic of China*, Richmond et al., accessed in draft on 10 May 2011 at <www.risk-and-regulation.com/wp-content/uploads/2010/07/Case-study-of-China.pdf>.

72. Park. *REACHing Asia Continued.* Social Science Research Network at <http://papers.ssrn.com/sol3/papers.cfm?abstract_id=1474504>.

73. SFDA Notice 146, 2006.

These devices are therefore subject to stricter control and safety oversight than other medical devices, including more detailed pre-market approval procedures.

7.5.3 Chemicals

Presently, chemical regulation in China follows the EU approach in that companies must apply to the Chemical Register Centre of the Ministry of Environmental Protection (CRC-MEP) for a registration certificate prior to manufacture/import of new chemicals. In this application process, companies must supply toxicological and other data to the CRC for all new chemical substances not listed on the *Inventory of Existing Chemical Substances Manufactured or Imported in China* (IECSC).[74] Registration requirements and enforcement guidelines are set out in the *Provisions on the Environmental Administration of New Chemical Substances*, in effect since October 2010[75] (Provisions), replacing the very much simpler Provisions of 2003.

The current Provisions are colloquially known as China-REACH, being modelled to a degree on the European system. Like the European REACH, China-REACH requires a local entity to take responsibility for the registration of new chemicals, giving rise to concerns that the technical and commercial information which must be provided for the purpose may be at risk in view of the irregular track record of protection of confidential information in China.[76] Further, the eco-toxicological testing must be done by an approved Chinese laboratory, which may cause delays as there are few OECD GLP-standard laboratories in China.

The Provisions' application to nanomaterials is not specifically stated, the question once again being whether a chemical substance already listed in bulk form must be re-registered if it shows different properties in nanoform. This may be implicit since the Provisions require information on hazard and exposure evaluation to be supplied in the risk assessment report, and this should entail different information where different properties affect the nature or degree of hazard or the exposure route.

Notified new chemical substances will be evaluated and categorized as either 'Regular' or 'Hazardous'. Within Hazardous Chemicals there is a subcategory named Critical Environmental Administrative Hazardous Substances (CEAHS). The CEAHS subcategory will include substances that are persistent, bio-accumulative and hazardous to ecological, environmental and human health. This subcategory is subject to more stringent regulatory oversight, including monitoring of environmental emissions during production and application process, additional measures for transportation and waste disposal.

---

74. A chemical substance is defined to be new if it is not on this list (Art. 3 of the Provisions).
75. MEP Decree No. 7 Available (in Chinese) at <www.zhb.gov.cn/gkml/hbb/bl/201002/t20100201_185231.htm>.
76. Feng, *China Promulgates Revised Environmental Administration Provisions for New Chemical Substances*, at <www.khlaw.com/showpublication.aspx?Show=3572>.

The information required depends upon the production or import volume, in bands of 1–10 tonnes, 10–100 tonnes, 100–1,000 and over 1,000 tonnes. But simplified notification is required even for volumes under 1 tonne. Actual volumes must be declared each year. A limited research and development information notice must be filed if the amount produced or imported is under 100 kg and is for eco-toxicological testing purposes. Production, import or use of an unregistered new chemical substance is banned.[77] Where the chemical in question is hazardous or of environmental concern, risk control measures are required and additional information must be included in the annual reporting.

Supervision is by the local environment protection department. If it finds that the production or processing activities using the new chemical substances caused or may cause instant or cumulative environmental pollution hazards immediate measures may be ordered to eliminate harmful or dangerous effects, and the situation must be reported to the Ministry. Under this provision, therefore, remedial action can in principle be ordered as soon as any adverse impact of the use of nanoforms of chemical substances becomes known even if the nanoform was not separately notifiable beforehand.

The fines for false declarations and other breaches have been increased to CNY 1,030,000. The registration will also be cancelled.

Measures overhauling China's regulatory regime for the manufacture, storage and transportation of hazardous substances will come into force on 1 December 2011 following the publication in March 2011 of a revised version of the Regulation on the Safe Management of Dangerous Chemicals.[78]

77. Article 5 of the Provisions.
78. State Council Decree 591.

Appendices

Appendix I
EU NanoCode

Brussels, 07/02/2008
C(2008) 424 final

COMMISSION RECOMMENDATION

of 07/02/2008

on a code of conduct for responsible nanosciences and nanotechnologies research

THE COMMISSION OF THE EUROPEAN COMMUNITIES,

Having regard to the Treaty establishing the European Community, and in particular Article 211 thereof,

Whereas:

(1) In its Communication to the Council, the European Parliament, the Economic and Social Committee and the Committee of the Regions "Towards a European research area" the Commission proposed in January 2000 the creation of a European Research Area[1] with a view to consolidating and structuring European research policy. In May 2007, in the Green Paper "The European Research Area: New Perspectives", the Commission re-launched a broad institutional and public debate on what should be done to create a unified and attractive European Research Area that would fulfil the needs and expectations of the scientific community, business and citizens.[2]

---

1. COM(2000) 6, 18.1.2000.
2. COM(2007) 161, 4.4.2007.

(2) The Commission adopted in February 2000 a Communication on the precautionary principle,[3] aiming to build a common understanding of how to assess, appraise, manage and communicate risks that science is not yet able to evaluate fully.

(3) In March 2000 the Lisbon European Council set for the Community the objective of becoming in the next decade the most competitive and dynamic knowledge economy in the world, capable of sustainable economic growth with more and better jobs and greater social cohesion.

(4) In 2004, with its Communication "Towards a European strategy for nanotechnology",[4] the Commission identified actions aimed at creating the Community added value necessary to remain competitive in this sector while ensuring its responsible development. In its conclusions of 24 September 2004,[5] the Council (Competitiveness) welcomed the proposed integrated, safe and responsible approach and the Commission's intention to draw up an Action Plan for nanotechnology.

(5) Taking into account the results of a public consultation, the Commission drew up in 2005 a Nanotechnologies Action Plan[6] which sets out coherent and interconnected actions for the immediate implementation of an integrated, safe and responsible strategy for nanosciences and nanotechnologies based on the priority areas identified in the Communication "Towards a European strategy for nanotechnology". Both Communications explicitly acknowledged that environmental, human health and safety aspects need to be integrated in all nanosciences and nanotechnologies research.

(6) Following the Nanosciences and nanotechnologies Action Plan, in January 2007 the European Group on Ethics in Science and New Technologies presented an Opinion on the ethical aspects of nanomedicine.[7]

(7) Following comments made during a public consultation on a previous opinion, the Scientific Committee on Emerging and Newly Identified Health Risks adopted in March 2006 a Modified Opinion on the appropriateness of existing methodologies to assess the potential risks associated with engineered and adventitious products of nanotechnologies.[8]

(8) In June 2006 the European Council adopted a revised sustainable development strategy fine-tuning the Community sustainable development strategy launched at the Gothenburg Summit in June 2001 that centred on objectives of environment and health protection and poverty eradication.

(9) In its conclusions[9] of 23 November 2007, the Council (Competitiveness) recognised the need to foster synergies and cooperation between all nanosciences and

3. COM(2000) 1, 2.2.2000.
4. COM(2004) 338, 12.5.2004.
5. Doc. 12487/4
6. COM(2005) 243, 7.6.2005.
7. EGE Opinion No 21, 17 January 2007.
8. SCENIHR/002/05, 10 March 2006.
9. Doc. 14865/07

nanotechnologies stakeholders, including the Member States, the Commission, academia, research centres, industry, financial bodies, non-governmental organisations and society at large.

(10) A first report on the implementation of the Nanotechnologies Action Plan for Europe was presented by the Commission in 2007.[10] In this report the Commission announced its intention to adopt a voluntary Code of Conduct for Responsible Nanosciences and Nanotechnologies Research.

(11) This Recommendation includes the Code of Conduct, aiming to promote integrated, safe and responsible nanosciences and nanotechnologies research in Europe for the benefit of society as a whole.

(12) The general principles and guidelines on actions to be taken outlined in this Recommendation benefited from a public consultation.

(13) This Recommendation provides Member States with an instrument to undertake further initiatives to ensure safe, ethical and sustainable nanosciences and nanotechnologies research in the European Union.

(14) This Recommendation also aims at contributing to proper coordination between Member States with a view to optimise synergies between all nanosciences and nanotechnologies research stakeholders at European and international levels,

HEREBY RECOMMENDS:

1. That Member States be guided by the general principles and guidelines for actions to be taken, set out in the Code of Conduct for Responsible Nanosciences and Nanotechnologies Research, in the Annex, as they formulate, adopt and implement their strategies for developing sustainable nanosciences and nanotechnologies (hereinafter N&N) research, in line with the Commission Nanotechnologies Strategy and Action Plan.

2. That Member States endeavour to follow these general principles and guidelines when implementing their national regulatory research and development strategies or developing sectoral and institutional research and development standards, taking into account pre-existing applicable N&N guidelines, good practices or regulations.

3. That Member States consider such general principles and guidelines on research to be an integral part of institutional quality assurance mechanisms by regarding them as a means for establishing funding criteria for national/regional funding schemes, as well as adopting them for the auditing, monitoring and evaluation processes of public bodies.

---

10. COM(2007) 505, 6.9.2007.

4. That Member States encourage the voluntary adoption of the Code of Conduct by relevant national and regional authorities, employers and research funding bodies, researchers, and any individual or civil society organisation involved or interested in N&N research and endeavour to undertake the necessary steps to ensure that they contribute to developing and maintaining a supportive research environment, conducive to the safe, ethical and effective development of the N&N potential.

5. That Member States cooperate with the Commission in order to review this recommendation every two years, as well as to monitor the extent to which relevant stakeholders have adopted and applied the Code of Conduct.

6. That the criteria for measuring such adherence to and application of the Code of Conduct be established and agreed with the Member States in relation to similar work undertaken at Community level.

7. That Member States, in their bilateral agreements on research strategies and activities with third countries and in their role as members of international organisations, take due account of this Recommendation when proposing research strategies and taking decisions, and duly coordinate with other Member States and the Commission.

8. That this Recommendation also be used as an instrument to encourage dialogue at all governance levels among policy makers, researchers, industry, ethics committees, civil society organisations and society at large with a view to increasing understanding and involvement by the general public in the development of new technologies.

9. That the Member States inform the Commission by 30 June 2008 and annually thereafter of any measures they have taken further to this Recommendation, inform it of the first results of its application and provide good practices.

Done at Brussels, 07/02/2008.

For the Commission
Janez POTOČNIK
Member of the Commission

ANNEX

CODE OF CONDUCT FOR RESPONSIBLE NANOSCIENCES AND NANOTECHNOLOGIES RESEARCH

This Code of Conduct provides Member States, employers, research funders, researchers and more generally all individuals and civil society organisations involved or interested in nanosciences and nanotechnologies (N&N) research ("all stakeholders") with guidelines favouring a responsible and open approach to N&N research in the Community.

The Code of Conduct is complementary to existing regulations. It does not limit or otherwise affect the possibilities of Member States to grant a wider measure of protection with regard to N&N research than is stipulated in this Code of Conduct.

Stakeholders who adhere to this Code of Conduct should also be inspired, where applicable, by the principles set out in the Charter of Fundamental Rights of the European Union.

The Code of Conduct will be regularly monitored and revised every two years by the Commission in order to take into account developments in N&N worldwide and their integration in European society.

1. SCOPE AND AIM

The Code of Conduct invites all stakeholders to act responsibly and cooperate with each other, in line with the N&N Strategy and Action Plan of the Commission, in order to ensure that N&N research is undertaken in the Community in a safe, ethical and effective framework, supporting sustainable economic, social and environmental development.

The Code of Conduct covers all N&N research activities undertaken in the European Research Area.

The Code of Conduct is voluntary. It offers a set of general principles and guidelines for actions to be taken by all N&N stakeholders. It should facilitate and underpin the regulatory and non-regulatory approaches outlined in the 2005-2009 N&N Action Plan for Europe, improving the implementation of current regulation and coping with scientific uncertainties.

The Code of Conduct should also be a European basis for dialogue with third countries and international organisations.

2. DEFINITIONS

For the purpose of the Code of Conduct, the following definitions apply:

a) Nano-objects: In the absence of recognised international terminology the generic term of 'nano-object' is used all throughout the Code of Conduct to designate products resulting from N&N research. It includes nanoparticles and their aggregation at nanoscale, nano-systems, nano-materials, nano-structured materials and nano- products.

b) N&N research: In the broadest sense understood here, N&N research encompasses all research activities dealing with matter at the nanometric scale (1 to 100 nm). It includes all man-made nano-objects be they engineered or involuntarily generated. Naturally occurring nano-objects are excluded from the scope of the Code of Conduct N&N research encompasses research activities from the most fundamental research to applied research, technology development and pre and co-normative research underpinning scientific advice, standards and regulations.
c) N&N stakeholders: Member States, employers, research funders, researchers and more generally all individuals and civil society organisations engaged, involved or interested in N&N research.
d) Civil society organisations: In the context of the Code of Conduct, civil society organisations are considered to be any legal entity that is non governmental, not-for- profit, not representing commercial interests, and pursuing a common purpose in the public interest.

3. GENERAL PRINCIPLES

This Code of Conduct is based on a set of general principles which call for actions aimed at guaranteeing their respect by all stakeholders.

3.1 Meaning

N&N research activities should be comprehensible to the public. They should respect fundamental rights and be conducted in the interest of the well-being of individuals and society in their design, implementation, dissemination and use.

3.2 Sustainability

N&N research activities should be safe, ethical and contribute to sustainable development serving the sustainability objectives of the Community as well as contributing to the United Nations' Millennium Development Goals.[11] They should not harm or create a biological, physical or moral threat to people, animals, plants or the environment, at present or in the future.

3.3 Precaution

N&N research activities should be conducted in accordance with the precautionary principle, anticipating potential environmental, health and safety impacts of N&N outcomes and taking due precautions, proportional to the level of protection, while encouraging progress for the benefit of society and the environment.

11. The United Nations Millennium Declaration, General Assembly resolution 55/2, 8.9.2000

3.4 Inclusiveness

Governance of N&N research activities should be guided by the principles of openness to all stakeholders, transparency and respect for the legitimate right of access to information. It should allow the participation in decision-making processes of all stakeholders involved in or concerned by N&N research activities.

3.5 Excellence

N&N research activities should meet the best scientific standards, including standards underpinning the integrity of research and standards relating to Good Laboratory Practices.[12]

3.6 Innovation

Governance of N&N research activities should encourage maximum creativity, flexibility and planning ability for innovation and growth.

3.7 Accountability

Researchers and research organisations should remain accountable for the social, environmental and human health impacts that their N&N research may impose on present and future generations.

4. GUIDELINES ON ACTIONS TO BE TAKEN

The guidelines set out in this point are based on the set of general principles described in point 3. They are meant to give guidance on how to achieve good governance, due respect for precaution, as well as wide dissemination and good monitoring of the Code of Conduct. The main responsibilities for action are indicated below, but all N&N stakeholders should contribute to their implementation as much as possible within the scope of their own remit.

4.1 Good governance of N&N research

Good governance of N&N research should take into account the need and desire of all stakeholders to be aware of the specific challenges and opportunities raised by N&N. A general culture of responsibility should be created in view of challenges and opportunities that may be raised in the future and that we cannot at present foresee.

4.1.1 Member States should cooperate with the Commission in order to maintain an open and pluralistic forum for discussion on N&N research at Community level as a means to stimulate the societal debate about N&N research, encouraging the identification and discussion of concerns and hopes and facilitating the emergence of possible initiatives and solutions. Accordingly, Member States should enhance communication on benefits, risks and uncertainties related to N&N research.

12. Directive 2004/9/EC and Directive 2004/10/EC

Specific attention should be paid to the younger and older members of the population.

4.1.2 With due respect for intellectual property rights, Member States, N&N research funding bodies, research organisations and researchers are encouraged to make easily accessible and understandable by lay people as well as by the scientific community all N&N scientific knowledge as well as related information such as relevant standards, references, labels, research on impacts, regulations and laws.

4.1.3 Member States should encourage private and public sector laboratories to share best practices in N&N research, with due respect for the protection of intellectual property.

4.1.4 N&N research organisations and researchers should ensure that scientific data and results are duly peer-reviewed before being widely disseminated outside the scientific community in order to ensure their clarity and balanced presentation.

4.1.5 Given its potential, Member States and N&N research organisations should ensure that N&N research is conducted at the highest level of scientific integrity. Questionable N&N research practices (not limited to plagiarism, falsification and fabrication of data) should be fought as they may entail risks for health, safety and the environment, raise public distrust and slow down the dissemination of benefits from research. Individuals signalling impropriety in research should be protected by their employers and national or regional laws.

4.1.6 Member States should ensure that appropriate human and financial resources are dedicated to the application of existing laws and regulations applicable to N&N research. Organisations performing N&N research activities should demonstrate transparently that they comply with relevant regulations.

4.1.7 National and local ethics committees and competent authorities should evaluate the manner of applying ethical review requirements to dual-use nanotechnology research. They should notably address the fundamental rights implications of any possible restrictions on informed consent and on publication of research results related to human health.

Favouring an inclusive approach

4.1.8 The broad directions of N&N research should be decided in an inclusive manner, allowing all stakeholders to enrich the preliminary discussions on these directions.

4.1.9 Member States, N&N research funding bodies, research organisations and researchers are encouraged to consider, at the earliest stages and through participatory foresight exercises, the future implications of technologies or objects being researched. This could allow the development of solutions to meet potential negative impacts caused by the use of a new object or technology at a later stage. Consultations with relevant ethics committees should be part of such foresight exercises as appropriate.

4.1.10 N&N research itself should be open to contributions from all stakeholders who should be informed and supported so that they can take an active part in the research activities, within the scope of their mission and mandate.

Key priorities

4.1.11 Research authorities and standardisation bodies should endeavour to adopt N&N standard terminology to facilitate the communication of scientific evidence. They should encourage standard measurement procedures as well as the use of appropriate reference materials in order to improve comparability of scientific data.

4.1.12 N&N research funding bodies should devote an appropriate part of N&N research to the development of methods and tools for risk assessment, the refinement of metrology at nano-scale and standardisation activities. In this context, particular attention should be paid to developing methods to assess the risk of second-generation, active nano-structures.

4.1.13 Member States, N&N research funding bodies and organisations should encourage fields of N&N research with the broadest possible positive impact. A priority should be given to research aiming to protect the public and the environment, consumers or workers and aiming to reduce, refine or replace animal experimentation.

4.1.14 N&N research funding bodies should carry out and publish balanced assessments, based on best available scientific data, of the potential costs, risks, and benefits of research areas eligible for funding.

Prohibition, restrictions or limitations

4.1.15 N&N research funding bodies should not fund research in areas which could involve the violation of fundamental rights or fundamental ethical principles, at either the research or development stages (e.g. artificial viruses with pathogenic potentials).

4.1.16 N&N research organisations should not undertake research aiming for non-therapeutic enhancement of human beings leading to addiction or solely for the illicit enhancement of the performance of the human body.

4.1.17 As long as risk assessment studies on long-term safety is not available, research involving deliberate intrusion of nano-objects into the human body, their inclusion in food (especially in food for babies), feed, toys, cosmetics and other products that may lead to exposure to humans and the environment, should be avoided.

4.2 Due respect for precaution

Given the deficit of knowledge of the environmental and health impacts of nano-objects, Member States should apply the precautionary principle in order to protect not only researchers, who will be the first to be in contact with nano-objects, but also professionals, consumers, citizens and the environment in the course of N&N research activities.

4.2.1 Students, researchers and research organisations involved in N&N research should take specific health, safety and environmental measures adapted to the particularities of the nano- objects manipulated. Specific guidelines on the prevention of pathologies induced by nano- objects should be developed in line with the Community Strategy 2007-2014 on Health and Safety at Work.[13]

4.2.2 N&N research organisations should apply existing good practices in terms of classification and labelling. In addition, as nano-objects might present specific properties due to their size, they should undertake research on systems (including e.g. the development of specific pictograms) aiming to inform researchers and more generally people likely to come into contact with nano-objects in research premises (e.g. security and emergency staff) so that they may take the necessary and appropriate protection measures in the course of their duties.

4.2.3 Public and private N&N research funding bodies should request that a risk assessment be presented along with each submission of a proposal for funding for N&N research.

4.2.4 N&N research funding bodies' programmes should include monitoring of the potential social, environmental and human health impacts of N&N over a relevant period of time.

Application of the precautionary principle should include reducing the gaps in scientific knowledge, and therefore undertaking further actions in research and development such as the following:

4.2.5 Research funding bodies should devote an appropriate part of N&N research to understanding the potential risks, notably to the environment and human health, induced by nano-objects, encompassing their whole life-cycle, from their creation up to their end of life, including recycling.

4.2.6 N&N research organisations and researchers should launch and coordinate specific N&N research activities in order to gain a better understanding of fundamental biological processes involved in the toxicology and ecotoxicology of nano-objects man-made or naturally occurring. They should widely publicise, when duly validated, data and findings on their biological effects, be they positive, negative or null.

4.2.7 N&N research funding bodies should launch and coordinate specific research activities in order to gain a better understanding of ethical, legal and societal impacts of the new fields opened by N&N. Information and communication technologies and biotechnology should receive particular attention as well as the convergence between these fields and cognitive sciences and N&N.

---

13. COM(2007) 62, 21.2.2007.

4.3 Wide dissemination and monitoring of the Code of Conduct

4.3.1 Member States should support the wide dissemination of this Code of Conduct, notably through national and regional public research funding bodies.

4.3.2 In addition to the existence of this Code of Conduct, N&N research funding bodies should make sure that N&N researchers are aware of all relevant legislation, as well as ethical and social frameworks.

4.3.3 As the application of the Code of Conduct should be monitored across the Community, Member States should cooperate with the Commission in order to devise adequate measures to carry out such monitoring at national level and guarantee synergies with other Member States.

Appendix II

Precautionary Matrix for Synthetic Nanomaterials

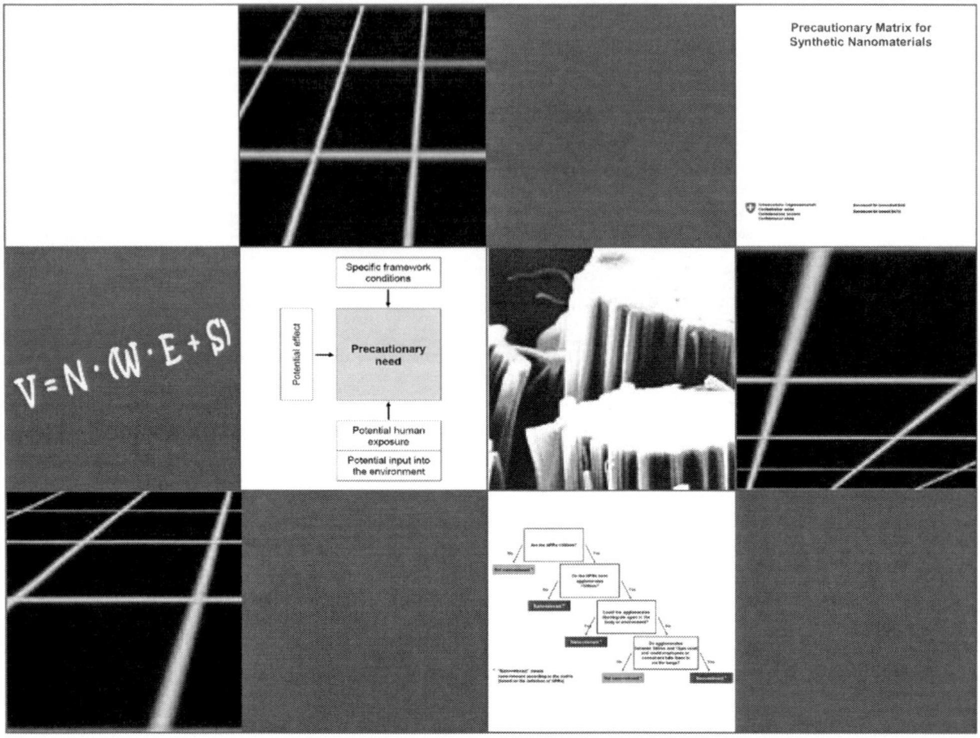

Version 2.1
14 July 2011

Schweizerische Eidgenossenschaft
Confédération suisse
Confederazione Svizzera
Confederaziun svizra

Federal Office of Public Health FOPH
Federal Office for the Environment FOEN

We would like to thank the following for their assistance in developing these Guidelines on the Precautionary Matrix for Synthetic Nanomaterials:

Federal Office for Agriculture
- Dr Katja Knauer

Federal Office of Public Health
- Dr Steffen Wengert
- Dr Martine Bourqui-Pittet
- Dr Christoph Studer (previously FOEN)

Federal Office for the Environment
- Dr Ernst Furrer
- Dr A. Weber

Eawag
- Dr Renata Behra
- Dr Beate Escher
- Prof. Kristin Schirmer

Empa, Inorganic Analytics
- Dr Andrea Ulrich

Empa, Materials-Biology Interactions
- Prof. Harald Krug
- Dr Peter Wick

Empa, Technology and Society
- Dr Bernd Nowack
- Claudia Som

EPFL, Powder Technology Laboratory
- Prof. Heinrich Hofmann

ETHZ, Functional Materials and Catalysis
- Dr Ludwig Limbach
- Prof. Jan Wendelin Stark

ETHZ, Safety and Environmental Technology
- Prof. Konrad Hungerbühler
- Christiane Lorenz
- Dr Natalie von Götz

Institute for Work and Health
- Dr Michael Riediker

State Secretariat for Economic Affairs
- Dr Livia Bergamin

SUVA
- Christoph Bosshard

Swissmedic
- Dr Beat Schmid

Independent Consultant for Risks and Insurance
- Dr Thomas K. Epprecht

University of Berne
- Prof. Peter Gehr

Project Director: TEMAS AG
- Dr Jürgen Höck
- Karl Höhener

Publishing details

Publisher
Federal Office of Public Health (FOPH)
Federal Office for the Environment (FOEN)

Citation
Höck J., Epprecht T., Furrer E., Hofmann H., Höhner K., Krug H., Lorenz C., Limbach L., Gehr P., Nowack B., Riediker M., Schirmer K., Schmid B., Som C., Stark W., Studer C., Ulrich A., von Götz N., Weber A., Wengert S., Wick P.: Precautionary Matrix for Synthetic Nanomaterials. Federal Office of Public Health and Federal Office for the Environment, Berne 2011, Version 2.1

Download PDF

http://www.bag.admin.ch/themen/chemikalien/00228/00510/05626/index.html?lang=en

FOPH, FOEN, TEMAS AG **Precautionary Matrix** confidential

A computerised version of the precautionary matrix exists which includes automatic evaluation of entries (www.nanotechnologie.admin.ch)

General informations

Version 2.1, 14 July 2011

Precautionary matrix completed by / responsible contact person		
Brief description of the considered nanospecific field (type of NPR, which surrounding, in which application)		
Brief description of the considered (process) step (production, packaging, transport, further stages of processing, disposal, use...), brief description		
Calculation of the precautionary need for employees?	O yes O no	
Calculation of the precautionary need for consumers?	O yes O no	
Calculation of the precautionary need for a specific disposal step?	O yes O no	Contact for information on disposal: see footer 28 in the guidelines
Are coated / functionalised NPRs involved?	O yes O no	if yes, see statements in the guidelines, section 4.5

It is recommended that the relevant accompanying documents be studied before using the precautionary matrix (see www.nanotechnologie.admin.ch)

A: relevant for workers
V: relevant for consumers
U: relevant for the environment
NPR: nanoparticles or nanorods

Please note: The precautionary matrix does not produce a definitive evaluation in terms of a risk assessment!

	Parameter classes	Parameter	Identifier	A	V	U	Appraisal	Basis for appraisal, applied measurement method
Nano-relevance	**Nano-relevance according to the precautionary matrix (i.e. contains NPR)**		N					
		Order of size of the primary particles (NPR) in the materials (free, bound, aggregated or agglomerated)	N1	x	x	x	O 1nm - 100nm, or 100nm - 500nm O over 500nm	
		Do the NPRs form agglomerates >500nm	N1a	x	x	x	O yes (proceed to N2) O no (or not known) -> review of nano-relevance concluded	
		Only if N1a = yes: In the body does deagglomeration of agglomerates (or aggregates) to primary NPRs or agglomerates <500nm occur	$N2_{A,V}$	x	x		O yes O no (proceed to N2a)	
		Only if N1a = yes: Under the respective environmental conditions does deagglomeration of agglomerates (or aggregates) to primary NPRs or agglomerates <500nm occur?	$N2_U$			x	O yes O no	
		Only if N2A,V = no: If agglomerates between 500nm and 10μm are present, can employees or consumers take them in through their lungs?	N2a	x	x		O yes O no	

FOPH, FOEN, TEMAS AG | **Precautionary Matrix** | confidential

It is recommended that the relevant accompanying documents be studied before using the precautionary matrix (see www.nanotechnologie.admin.ch)

A: relevant for workers
V: relevant for consumers
U: relevant for the environment
NPR: nanoparticles or nanorods

Please note:
The precautionary matrix does not produce a definitive evaluation in terms of a risk assessment!

	Parameter classes	Parameter	Identifier	A	V	U	Appraisal	Basis for appraisal, applied measurement method
Specific framework conditions	**Specific framework conditions on the information status**		S					
		Is the origin of the (nanoscale) starting materials known?	S1	x	x	x	O yes O partly O no	
		Is sufficient information available to complete the precautionary matrix for nanoscale starting materials?	S2	x	x	x	O yes O partly O no	
		Are the subsequent users of the considered NPRs known?	S3	x	x	x	O yes O partly O no	
		How accurately is the material system known, or can disturbing factors (e.g. impurities) be estimated?	S4	x	x	x	O accurately O not accurately O unknown	
Potential effect	**Potential effect**		W					
		Redox activity and/or catalytic activity of the NPRs present in the nanomaterial	W1	x	x	x	O low O medium O high	
		Stability (half-life) of the NPRs present in the nanomaterial in the body	$W2_{A,V}$	x	x		O hours O days-weeks O months	
		Stability (half-life) of the NPRs present in the nanomaterial under environmental conditions	$W2_U$			x	O hours O days-weeks O months	

Appendix II

It is recommended that the relevant accompanying documents be studied before using the precautionary matrix (see www.nanotechnologie.admin.ch)

A: relevant for workers
V: relevant for consumers
U: relevant for the environment
NPR: nanoparticles or nanorods

Please note.
The precautionary matrix does not produce a definitive evaluation in terms of a risk assessment!

Potential exposure of humans / potential input into the environment

Parameter classes	Parameter	Identifier	A	V	U	Appraisal	Basis for appraisal, applied measurement method
Physical surroundings		E1					
	Air	E1.1	x	x	x	O yes O no	
	Aerosols <10 µm	E1.2	x	x	x	O yes O no	
	Aerosols >10 µm	E1.3	x	x	x	O yes O no	
	Liquid media	E1.4	x	x	x	O yes O no	
	Solid matrix, not stable under conditions of use	E1.5	x	x	x	O yes O no	
	Solid matrix, stable under conditions of use, NPR mobile	E1.6	x	x	x	O yes O no	
	Solid matrix, stable under conditions of use, NPR not mobile	E1.7	x	x	x	O yes O no	
Maximum possible exposure of humans		E2					
	Amount of NPR which a worker handles per day	E2.1	x			O up to 1.2mg O 1.2mg - 12mg O over 120mg	
	Amount of NPR with which a worker comes into contact in the "worst case"	E2.2	x			O up to 12mg O 12mg - 120mg O over 120mg	
	Frequency with which a worker handles the NPR(s)	E2.3	x			O monthly O weekly O daily	
	Amount of NPR which a consumer handles daily through the utility product	E2.4		x		O up to 1.2mg O 1.2mg - 12mg O over 12mg	
	Frequency with which a consumer uses the utility product	E2.5		x		O monthly O weekly O daily	
Maximum possible input into the environment		E3					
	Amount of NPR reaching the environment from wastewater, exhaust gases, solid waste per year	E3.1			x	O up to 5kg O 5kg - 500kg O over 500kg	
	Amount of NPR in utility products per year	E3.2				O up to 5kg O 5kg - 500kg O over 500kg	
	Amount of disposed NPR per year	E3.3			x	O up to 5kg O 5kg - 500kg O over 500kg	

Appendix III

Final Version of Classification, Labelling and Packaging of Nanomaterials in REACH and CLP

EUROPEAN COMMISSION

ENVIRONMENT DIRECTORATE-GENERAL
Water, Chemicals & Biotechnology
Chemicals and Nanomaterials

ENTERPRISE AND INDUSTRY DIRECTORATE-GENERAL
Chemicals, Metals, Forest-based & Textile Industries
REACH

Brussels, 3 December 2009
Doc.CA/90/2009 Rev2

Classification, labelling and packaging of nanomaterials in REACH and CLP

This document reflects the current state of ongoing discussions within the REACH-CLP Competent Authorities and its subgroup on nanomaterials on how CLP and REACH applies to nanomaterials. Further updates of this document can be expected following continued discussions. Stakeholders are invited to take note of the content of this document and follow its further development.

Appendix III

CLASSIFICATION, LABELLING AND PACKAGING OF NANOMATERIALS

1. INTRODUCTION

The aim of this paper is to describe some relevant aspects known at this stage related to the classification, labelling and packaging of nanomaterials in accordance with REACH and CLP[1] and how registrants should proceed with this issue in order to ensure a coherent approach. The terminology applied in this paper is in line with that of CA/59/2008.[2] The approach proposed below is in line with the guidance on Classification, Packaging and Labelling in the context of REACH and CLP which were finalised in summer 2009.

2. CLASSIFICATION AND LABELLING OF NANOMATERIALS

2.1. Regulatory process

The legislation on classification and labelling (67/548/EEC and 1999/45/EC),)[3] as well as the new Classification, Labelling and Packaging (CLP) Regulation[4] (1272/2008/EC) which came into force January 20, 2009, implementing the Globally Harmonised Systems (GHS), also provides the general framework for the classification and labelling of nanomaterials. According to the transitional arrangements, substances must be classified based on 67/548/EEC until 30 Nov 2010 and the information about the old classification has to be provided in the SDS until 1 June 2015. Mixtures[5] must be classified based on the old legislation until 31 May 2015. If preferred, companies can apply CLP earlier on a voluntary basis.

The purpose of the CLP regulation is to ensure a high level of protection of human health and the environment as well as the free movement of substances, mixtures and certain articles. Manufacturers, importers and downstream users are obliged to classify substances and mixtures (preparations) which are placed on the market. Suppliers are obliged to label and package substances and mixtures which are placed on the market. Substances that are placed on the market and either, meet

1. CLP Regulation has now replaced the Title XI in REACH and other issues related to classification and labelling are further modified in CLP.
2. <http://ec.europa.eu/enterprise/sectors/chemicals/files/reach/nanomaterials_en.pdf>; <http://ec.europa.eu/environment/chemicals/reach/pdf/nanomaterials.pdf>.
3. In force.
4. Regulation of the European Parliament and of the Council on classification, labelling and packaging of substances and mixtures, amending and repealing Directives 67/548/EEC and 1999/45/EC, and amending Regulation (EC) No 1907/2006, entered in force in January 2009. After a transitional period, the new regulation will replace the current rules on classification, labelling and packaging of substances (Directive 67/548/EEC) and mixtures (Directive 1999/45/EC). The rules on harmonised classification and notification to the inventory will apply from entry into force. The deadline for substance classification according to the new rules will be 1 December 2010 and for mixtures 1 June 2015.
5. The term 'mixture' in this document refers to 'preparation' under REACH.

the criteria for classification as hazardous or, are subject to registration in accordance with REACH, have to be notified to the European Chemicals Agency (ECHA), in general by 3 January 2011.

ECHA will establish a CLP Inventory, containing the information provided in accordance with Article 40(1) of CLP. This information, as far as it corresponds to the information specified in Article 119 (1) of REACH, will be made publicly available. The list of harmonised classifications of substances at Community level (CLP, Part 3 of Annex VI) will continue to be updated. The procedure for such harmonisation includes a dossier made by Member State CAs (revisions and new proposals) or manufacturers, importers or downstream users (new proposals) an opinion of the Risk Assessment Committee of ECHA and a decision of the Commission that shall be adopted by comitology procedure (regulatory procedure with scrutiny).

Information on classification and labelling must be included in the respective Safety Data Sheets (SDSs) of the substance and mixture when communicating throughout the supply chain and in the registration dossier, if appropriate.

2.2. **Classification and labelling of nanomaterials**

"When evaluating the available information for the purpose of classification, the manufacturers, importers and downstream users shall consider the forms or physical states in which the substance or mixture is placed on the market and in which it can reasonably be expected to be used" (CLP Art. 9(5)). The hazard classification should be based on *available data* on the intrinsic properties that relate to the forms or physical state in which the substance or mixture is placed on the market and in which it can reasonably be expected to be used[6] (CLP Art.5(1), 6(1) and 8(6)). The group of companies[7] or individual companies are expected to make use of relevant available information created e.g. under REACH and conduct additional testing where required for physico-chemical properties (CLP Art.8(1-2)).

According to the guidance on classification,[8] putative forms may comprise properties such as crystal structure, particle size, homogeneity (e.g., emulsion) and texture (e.g., viscosity, tablet form). Examples of physical state factors are agglomeration state, surface treatment, moisture content, residual solvent, activation or stabilisation. Any tested sample should be representative for the substance or mixture as it is placed on the market. Therefore if the form or physical state of a substance is changed, it has to be evaluated whether that affects the classification.

For example, a substance with different particle sizes or forms can have different classifications, as is the case for e.g. nickel and nickel powder (particle diameter < 1 mm). If substances are produced/imported both at nanoscale and as

6. In this context, reasonably expected use summarises all physical forms and states of a substance or mixture that may occur during intended or unintended use conditions.
7. For instance the respective Substance Information Exchange Forum (SIEF) of the substance.
8. <http://guidance.echa.europa.eu/docs/guidance_document/clp_en.htm?time¼1256044522>.

bulk, a separate classification and labelling may be required if the available data on the intrinsic properties indicates a difference in hazard class between the nano form and ones the bulk form.[9]

Due to the knowledge gaps on the properties of the rapidly emerging nanomaterials, it is not yet possible to identify any systematic rules for the toxicological characteristics of all nanomaterials (SCENIHR 2006 and 2007). The SCENIHR opinion (2009) recognised, with reference to the discussions in the OECD and ISO that a consensus is emerging on the physical-chemical properties of nanoparticles that need to be addressed in the risk assessment process of nanomaterials. The main physical parameters of interest with respect to nanoparticle characterisation are: the size, shape, specific surface area, aspect ratio, agglomeration/aggregation state, and size distribution, and surface morphology/topography, structure including crystallinity and defect structure and solubility. The main chemical parameters are: structural formula/molecular formula, composition of nanomaterials (including degree of purity, known impurities or additives), phase identity, surface chemistry, composition, charge tension, reactive sites, physical structure, photocatalytic properties, zeta potential and hydrophilicity/lipophilicity (SCENIHR, 2009).

Moreover, the SCENIHR stated that not all nanoparticle formulations have been found to induce a more pronounced hazard than the bulk formulations of the same substance. This suggests that the hazard characterisation of nanoparticle formulations should be carried out on a case-by-case basis. The potential influence of several physico-chemical parameters for the hazard profile of nanomaterials needs to be considered in the hazard assessment.

Another important question, which is scrutinised by the OECD Working Party on Manufactured Nanomaterials (WPMN), is the adequacy of current test guidelines to deliver results for hazard classification of nanomaterials. Important issues in relation to the adequacy of the test guidelines remain unanswered. So far it has been concluded in the preliminary ~~initial~~ review that *"Many of the OECD Test Guidelines are applicable, with conditions in some cases, while some are inadequate for testing MN as measuring, dosing, delivery and tracking nanomaterials are not reliably accomplished at this stage. Therefore, the review of OECD Test Guidelines reinforced the need for a guidance document(s) for sample preparation and dosimetry.*[10] A special guidance addressing the latter issue has been prepared with priority and is currently about to be presented for declassification.

France has raised several issues with UN GHS on the classification of nanomaterials (ST/SG/AC.10/C.4/2009/3). In case further work will be needed, such initiative should preferably be taken under the umbrella of the UN SCE GHS (UN Subcommittee of Experts on GHS) to enable a harmonised approach at the global level.

Due to above-mentioned reasons, the classification and labelling of nanomaterials has to be done on a case-by-case basis.

---

9. REACH Article 10(a)
10. <http://www.olis.oecd.org/olis/2009doc.nsf/LinkTo/NT000049AE/$FILE/JT03267900.PDF>.

3. CLASSIFICATION AND LABELLING OF NANOMATERIALS BY REGISTRANTS

The task of registrants is to facilitate data sharing for registration and agree on the classification and labelling of the substances concerned. The classification and labelling of nanomaterials should follow the rules set out in CLP, but the experience with nanomaterials is still very limited and the approach described below represent only some general elements. Taking into account the current knowledge on the properties of nanomaterials, a proper substance identification[11] is essential and it is advisable that registrants would consider the following approaches in the classification and labelling of nanomaterials:

1) The data sharing, should cover all relevant information including (but not limited to) sizes, forms and morphologies of the nanomaterial;
2) It is vital to evaluate whether changes in e.g. size, form or physical state influence considerably on the hazardous properties of the nanomaterial;
3) All available information of nanomaterials should be evaluated in the hazard assessment of the nanomaterial (REACH, Article 1(3));
4) Special attention needs to be devoted to the appropriateness of the sample preparation and dosimetry used in the testing of nanomaterials,
5) The classification of nanomaterials should be done on a case-by-case basis giving due consideration to relevant available data on e.g. the bulk form and read-across to other nanomaterials.
6) On the basis of the classification in accordance with CLP, the nanomaterial should also be labelled and packaged in accordance to CLP.

11. As indicated in REACH Annex VI, item 2.

Appendix IV

Commission Recommendation on the Definition of Nanomaterial

COMMISSION RECOMMENDATION

of 18 October 2011

on the definition of nanomaterial

(Text with EEA relevance)

(2011/696/EU)

THE EUROPEAN COMMISSION,

Having regard to the Treaty on the Functioning of the European Union, and in particular Article 292 thereof,

Whereas:

(1) The Commission Communication of 7 June 2005 'Nanosciences and nanotechnologies: An action plan for Europe 2005–2009'[1] defines a series of articulated and interconnected actions for the immediate implementation of a safe, integrated and responsible approach for nanosciences and nanotechnologies.

(2) The Commission, in line with the commitments made in the Action Plan, carefully reviewed relevant Union legislation with a view to determine the applicability of the existing regulations to the potential risks of nanomaterials. The result of the review was contained in the Commission Communication of 17 June 2008 'Regulatory aspects of nanomaterials'.[2] The Communication concluded

---

1. COM(2005) 243 final.
2. COM(2008) 366 final.

that the term 'nanomaterials' is not mentioned specifically in Union legislation but that existing legislation in principle covers the potential health, safety and environmental risks in relation to nanomaterials.

(3) The European Parliament in its resolution of 24 April 2009 on regulatory aspects of nanomaterials[3] called, inter alia, for the introduction of a comprehensive science-based definition of nanomaterials in Union legislation.

(4) The definition in this Recommendation should be used as a reference for determining whether a material should be considered as a 'nanomaterial' for legislative and policy purposes in the Union. The definition of the term 'nanomaterial' in Union legislation should be based solely on the size of the constituent particles of a material, without regard to hazard or risk. This definition, based only on the size of a material, covers natural, incidental or manufactured materials.

(5) The definition of the term 'nanomaterial' should be based on available scientific knowledge.

(6) Measuring size and size distributions in nanomaterials is challenging in many cases and different measurement methods may not provide comparable results. Harmonised measurement methods must be developed with a view to ensuring that the application of the definition leads to consistent results across materials and over time. Until harmonised measurement methods are available, best available alternative methods should be applied.

(7) The European Commission Joint Research Centre Reference Report 'Considerations on a Definition of Nanomaterial for Regulatory purposes'[4] suggests that a definition of nanomaterials should address particulate nanomaterials, be broadly applicable in Union legislation and be in line with other approaches worldwide. Size should be the only defining property which necessitates a clear definition of the nanoscale limits.

(8) The Commission mandated the Scientific Committee on Emerging and Newly Identified Health Risks (SCENIHR) to provide scientific input on elements to consider when developing a definition of the term 'nanomaterial' for regulatory purposes. The opinion 'Scientific basis for the definition of the term "Nanomaterial"' was subject to a public consultation in 2010. In its opinion of 8 December 2010,[5] SCENIHR concluded that size is universally applicable to nanomaterials and the most suitable measurand. A defined size range would facilitate a uniform interpretation. The lower limit was proposed at 1 nm. An upper limit of 100 nm is commonly used by general consensus, but there is no scientific evidence to support the appropriateness of this value. The use of a single upper limit value might be too limiting for the classification of nanomaterials and a differentiated approach might be more appropriate. For regulatory purposes, the number size distribution should

3. P6_TA(2009) 0328.
4. EUR 24403 EN, June 2010.
5. <http://ec.europa.eu/health/scientific_committees/emerging/docs/ scenihr_o_032.pdf>.

also be considered using the mean size and the standard deviation of the size to refine the definition. The size distribution of a material should be presented as size distribution based on the number concentration (i.e. the number of objects within a given size range divided by the number of objects in total) and not on the mass fraction of nanoscale particles in the nanomaterial as a small mass fraction may contain the largest number of particles. SCENIHR identified certain specific cases where the application of the definition can be facilitated by using the volume specific surface area as proxy to determine if a material falls within the defined nano size range.

(9) The International Organisation for Standardisation defines the term ‘nanomaterial’ as ‘material with any external dimensions in the nanoscale or having internal structure or surface structure in the nanoscale’. The term ‘nanoscale’ is defined as size range from approximately 1 nm to 100 nm.[6]

(10) The number size distribution should cover for the fact that nanomaterials most typically consist of many particles present in different sizes in a particular distribution. Without specifying the number size distribution, it would be difficult to determine if a specific material complies with the definition where some particles are below 100 nm while others are not. This approach is in line with the opinion of SCENIHR that the particle distribution of a material should be presented as the distribution based on the number concentration (i.e. the particle number).

(11) There is no unequivocal scientific basis to suggest a specific value for the size distribution below which materials containing particles in the size range 1 nm–100 nm are not expected to exhibit properties specific to nanomaterials. The scientific advice was to use a statistical approach based on standard deviation with a threshold value of 0,15%. Given the widespread occurrence of materials that would be covered by such a threshold and the need to tailor the scope of the definition for use in a regulatory context, the threshold should be higher. A nanomaterial as defined in this recommendation should consist for 50% or more of particles having a size between 1 nm–100 nm. In accordance with SCENIHR’s advice, even a small number of particles in the range between 1 nm–100 nm may in certain cases justify a targeted assessment. However, it would be misleading to categorise such materials as nanomaterials. Nevertheless there may be specific legislative cases where concerns for the environment, health, safety or competitiveness warrant the application of a threshold below 50%.

(12) Agglomerated or aggregated particles may exhibit the same properties as the unbound particles. Moreover, there can be cases during the life-cycle of a nanomaterial where the particles are released from the agglomerates or aggregates. The definition in this Recommendation should therefore also include particles in agglomerates or aggregates whenever the constituent particles are in the size range 1 nm–100 nm.

---

6. <http://cdb.iso.org>.

(13) At present it is possible to measure the specific surface area by volume for dry solid materials or powders with the nitrogen adsorption method ('BET-method'). In those cases the specific surface area can be used as a proxy to identify a potential nanomaterial. New scientific knowledge may expand the possibility to use this and other methods to other types of materials in the future. There can be a discrepancy between the measurement of the specific surface area and the number size distribution from one material to another. Therefore it should be specified that results for number size distribution should prevail and it should not be possible to use the specific surface area to demonstrate that a material is not a nanomaterial.

(14) Technological development and scientific progress continue with great speed. The definition including descriptors should therefore be subject to a review by December 2014 to ensure that it corresponds to the needs. In particular, the review should assess whether the number size distribution threshold of 50% should be increased or decreased and whether to include materials with internal structure or surface structure in the nanoscale such as complex nano-component nanomaterials including nano-porous and nano-composite materials that are used in some sectors.

(15) Guidance and standardised measurement methods as well as knowledge about typical concentrations of nanoparticles in representative sets of materials should be developed where feasible and reliable to facilitate the application of the definition in a specific legislative context.

(16) The definition set out in this Recommendation should not prejudge nor reflect the scope of application of any piece of Union legislation or of any provisions potentially establishing additional requirements for those materials, including those relating to risk management. It may in some cases be necessary to exclude certain materials from the scope of application of specific legislation or legislative provisions even if they fall within the definition. It may likewise be necessary to include additional materials, such as some materials with a size smaller than 1 nm or greater than 100 nm in the scope of application of specific legislation or legislative provisions suited for a nanomaterial.

(17) Given the special circumstances prevailing in the pharmaceutical sector and the specialised nano-structured systems already in use, the definition in this Recommendation should not prejudice the use of the term 'nano' when defining certain pharmaceuticals and medical devices,

HAS ADOPTED THIS RECOMMENDATION

1. Member States, the Union agencies and economic operators are invited to use the following definition of the term 'nanomaterial' in the adoption and implementation of legislation and policy and research programmes concerning products of nanotechnologies.

2. 'Nanomaterial' means a natural, incidental or manufactured material containing particles, in an unbound state or as an aggregate or as an agglomerate and where,

for 50% or more of the particles in the number size distribution, one or more external dimensions is in the size range 1 nm–100 nm.

In specific cases and where warranted by concerns for the environment, health, safety or competitiveness the number size distribution threshold of 50% may be replaced by a threshold between 1 and 50%.

3. By derogation from point 2, fullerenes, graphene flakes and single wall carbon nanotubes with one or more external dimensions below 1 nm should be considered as nanomaterials.

4. For the purposes of point 2, 'particle', 'agglomerate' and 'aggregate' are defined as follows:

(a) 'particle' means a minute piece of matter with defined physical boundaries;
(b) 'agglomerate' means a collection of weakly bound particles or aggregates where the resulting external surface area is similar to the sum of the surface areas of the individual components;
(c) 'aggregate' means a particle comprising of strongly bound or fused particles.

5. Where technically feasible and requested in specific legislation, compliance with the definition in point 2 may be determined on the basis of the specific surface area by volume. A material should be considered as falling under the definition in point 2 where the specific surface area by volume of the material is greater than 60 m^2/cm^3. However, a material which, based on its number size distribution, is a nanomaterial should be considered as complying with the definition in point 2 even if the material has a specific surface area lower than 60 m^2/cm^3.

6. By December 2014, the definition set out in points 1 to 5 will be reviewed in the light of experience and of scientific and technological developments. The review should particularly focus on whether the number size distribution threshold of 50% should be increased or decreased.

7. This Recommendation is addressed to the Member States, Union agencies and economic operators.

Done at Brussels, 18 October 2011.

For the Commission
Janez POTOČNIK
Member of the Commission

Bibliography

Books

Drexler, E. *Engines of Creation: The Coming Era of Nanotechnology* (Anchor Books, 1986).

Ellul, J. *The Technological Society*. Trans. Wilkinson (Alfred A. Knopf, 1964).

Lintner, K. (ed.). *Global Regulatory Issues for the Cosmetics Industry* (William Andrew Inc., 2009).

Murashov, V. and Howard, J. *Nanotechnology Standards* (Springer, 2011).

Papers

Ashwood et al., *Fine Particles That Adsorb Lipopolysaccharides via Bridging Calcium Cations May Mimic Bacterial Pathenogenicity towards Cells*. Exp Biol Med 232(1) (2007): 107–117.

Feng et al., *Functionalized Europium Oxide Nanoparticles Used as a Fluorescent Label in an Immunoassay for Atrazine*. Anal. Chem. 75 (2003): 5282–5286.

Feynman, R. *There's Plenty of Room at the Bottom*. Caltech Engineering and Science, 23(5) (February 1990): 22–36.

Handy et al. *The Ecotoxicology and Chemistry of Manufactured Nanoparticles*. Ecotoxicology 17 (2008): 287–314.

Iijima et al. *Single-Shell Carbon Nanotubes of 1-nm Diameter*. Nature 363(6430) (1993): 603–605.

Jones et al. *RepRap – The Replicating Rapid Prototyper*. Robotica 29 (2011): 177–191.

Joy. *Why the Future Doesn't Need us*. Wired (April 2000).

Kahru and Dubourguier. *Review: From Ecotoxicology to Nanoecotoxicology*. Toxicology 269 (2010): 105–119.

Klaine et al. *Nanomaterials in the Environment: Behavior, Fate, Bioavailability, and Effects*. Environ. Toxicol. Chem. 27 (2008): 1825–1851.

Ludlow. *The Readiness of Australian Food Regulations for the Use of Nanotechnology in Food and Food Packaging*. UMonashLRS 14 (2009).

Manning et al. *Diseases Caused by Asbestos: Mechanisms of Injury and Disease Development*. Int. Immunopharmacol 2 (2–3) (February 2002): 191–200.

Maynard, A. *Don't Define Nanomaterials*. Nature July 2011, 475 at 31.

Park. *REACHing Asia Continued*. Social Science Research Network, September 2009, at <http://papers.ssrn.com/sol3/papers.cfm?abstract_id=1474504>.

Poland et al. *Carbon Nanotubes Introduced into the Abdominal Cavity of Mice Show Asbestos-Like Pathogenicity in a Pilot Study*. Nature Nanotechnology 3 (2008): 423.

Richmond et al. *Mapping Emerging Nanotechnology Policies and Regulations: The Case of the People's Republic of China*. Accessed in draft on 10 May 2011 at <www.risk-and-regulation.com/wp-content/uploads/2010/07/Case-study-of-China.pdf>.

Seaton et al. *Nanoparticles, Human Health Hazard and Regulation*. J.R. Soc. Interface 7 (2010): S119–S129.

Song, Li and Du. *Exposure to Nanoparticles Is Related to Pleural Effusion, Pulmonary Fibrosis and Granuloma*. Eur Respir J 34 (2009): 559–567.

Toumey. *Reading Feynman into Nanotechnology: A Text for a New Science*. Techné 13(3) (2008): 133–168.

Toyokuni. *Mechanisms of Asbestos-Induced Carcinogenesis*. Nagoya J Med Sci. 71(1–2) (February 2009): 1–10.

Zykov et al. *Self-reproducing Machines*. Nature 435(7038) (2005): 163–164.

Table of Cases

EPO Decisions

G2/88 *MOBIL OIL/Friction-reducing additives* Enlarged Board of Appeal of the EPO, 35
T-0206/83 *ICI/pyridine herbicides* [1986] 5 EPOR 232, 38
T-0226/85 *UNILEVER/Stable bleaches* [1988] OJ EPO 336, 42
T-0396/89 *UNION CARBIDE/ high tear strength polymer* [1992] EPOR 312, 39
G-01/92 *Availability to the Public* [1993] EPOR 241, 38
T-0929/92/*AGREVO/Triazoles* [1996] EPOR 171, 42
G-02/98 *Entitlement to claim priority* [2001] OJ EPO 413, 37
T-1121/03/*UNION CARBIDE/Indicator ligands* [2006] EPOR 4, 42

English Cases

Donoghue v. Stevenson [1932] AC 562, 84
Evans v. Triplex Safety Glass Co Ltd [1936] 1 All ER 283, 84
Roe v. Minister of Health [1954] 2 QB 66, 84
Haley v. London Electricity Board [1965] AC 778, 85
Vacwel Engineering Co Ltd v. BDH Chemicals Ltd [1971] 1 QB 88, 84
Dunford & Elliot Ltd v. Johnson & Firth Brown [1978] FSR 143, 54
Thomas Marshall (Exports) Ltd v. Guinle [1978] 3 WLR 116, [1979] Ch 227, 54
Genentech's Patent [1987] RPC 553; [1989] RPC 147, 35
Lansing Linde Ltd v. Kerr [1991] 1 All ER 418, 52
Hollis v. Dow Corning (1996) 129 DLR (4th) 609, 84
Gerber Garment Technology Inc v. Lectra Systems Ltd [1997] RPC 269, 55
Shelley Films Ltd v. Rex Features Ltd [1994] EMLR 134 *Kirin-Amgen v. TKT* [2005] RPC 169, 44

SmithKlineBeecham v. Apotex Europe [2004] EWCA Civ 1568; *Les Laboratoires Servier v. Apotex Inc* [2008] EWCA Civ 445, 36
Nokia Corporation v. Interdigital Technology Corporation [2005] EWCA Civ 614, 64
Monsanto Technology v. Cargill International [2007] EWHC 2257 (Pat), 45
Abbott Laboratories Inc v. Evysio Medical Devices ULC [2008] EWHC 800, 40
Alexis v. Newham LBC [2009] ICR 1517, 84
Oxonica v. Neuftec [2009] EWCA Civ 668, 68
Vestergaard Frandsen A/S and Ots v. Bestnet Europe Ltd and Ots ([2009] EWHC (Ch), 54
OB v. Aventis Pasteur SA [2010] UKSC 23, 75

European Union Cases

Centrafarm v. Sterling Drug ([1974] ECR 1147), 50
Agreements on Video Cassette Recorders, Re (C78/156) [1978] 2 CMLR 160, 60
Volvo v. Erik Veng (UK) Ltd (Case 238/87) [1988] ECR 2605, 82
CICRA v. Renault (Case 53/87) [1988] ECR 6039, 82
Radio Telefis Eireann (RTE) and Independent Television Publications Ltd (ITP) v. Commission (Magill) (Joined cases C-241/91 and C-242/91) [1995] ECR
1-743, 82
European Commission v. UK [1997] All ER (EC) 481, 76
MPEG-2 Licensing Programme 17/07/98 (Commission Decision IV/C-3/36.849), 61
Pfizer Animal Health SA v. Council and Alpharma Inc. v. Council (Cases T-13/99 and T-70/99), 107
OB v. Aventis Pasteur MSD and Aventis Pasteur SA (Case C-127/04) [2006] 1 WLR 1606, 75
Microsoft v. Commission, (Case T-201/04) [2007] ECR II-1491, 82
Monsanto Technology v. Cefetra BV (Case C-428/08), 45
O'Byrne v. Aventis Pasteur SA (Case C-358/08), 74, 75
R (on the application of SPCM SA, CH Erbsloh KG, Lake Chemicals and Minerals Ltd, Hercules Inc.) v. Secretary of State for the Environment, Food and Rural Affairs (Case C-558/07), ECJ decision 7.7.09, 137

US Cases

Roche Products v. Bolar Pharmaceutical, 733 F.2d 858 (Fed. Cir. 1984), 26
Rambus Inc. v. FTC, 522 F.3d 456, 466–67 (D.C. Cir. 2008), *cert. denied*, 129 S. Ct. 1318 (2009), 63

WTO Dispute Panel Decisions

EC – measures concerning meat and meat products (Hormones) Report of the Appellate Body AB-1997-4, 105

Dutch Cases

Table of Legislation

1. International treaties

2. European Treaties and legislation

3. National legislation

Index